Biotechnology and Environmental Science

Molecular Approaches

Biotechnology and Environmental Science

Molecular Approaches

Edited by

S. Mongkolsuk
Chulabhorn Research Institute
Bangkok, Thailand

P. S. Lovett
University of Maryland
Catonsville, Maryland

and

J. E. Trempy
Oregon State University
Corvallis, Oregon

Springer Science+Business Media, LLC

Library of Congress Cataloging in Publication Data

Biotechnology and environmental science: molecular approaches / edited by S. Mongkolsuk, P.S. Lovett, and J. E. Trempy.
    p.   cm.
    "Proceedings of an international conference on biotechnology and environmental science: molecular approaches, held August 21–24, 1990, in Bangkok, Thailand" — T.p. verso.
    Includes bibliographical references and index.

    1. Biotechnology — Congresses. 2. Bioremediation — Congresses. 3. Agricultural biotechnology — Congresses. 4. Microbial biotechnology — Congresses. I. Mongkolsuk, S. II. Lovett, P. S. III. Trempy, J. E.

TP248.14.B5592   1993   92-40429
660'.6 — dc20   CIP

Proceedings of an International Conference on Biotechnology and Environmental Science: Molecular Approaches (BESMA), held August 21–24, 1990, in Bangkok, Thailand

ISBN 978-1-4757-6990-6   ISBN 978-0-585-32386-2 (eBook)
DOI 10.1007/978-0-585-32386-2

© 1992 Springer Science+Business Media New York
Originally published by Plenum Press, New York in 1992.
Softcover reprint of the hardcover 1st edition 1992

FOREWORD

Biotechnology and environmental science are two of the most rapidly developing and exciting fields. Both areas have contributed to improving the standard of living for mankind. These areas of research offer opportunities to scientists to seeing the basic research in their laboratories being translated into useful applications, scale up processes and commercial possibilities in a relatively short time. Globally, many nations have invested both human and financial resources to facilitating the development of biotechnology and environmental science with the aim of improving the environment, quality of life and commercialization. These technologies are needed in the developing countries where the proper application of both technologies would not only result in an economic growth but also a cleaner environment. Thus, there is an urgent need for these countries to be aware of the tremendous possibilities these technologies have to offer. The Chulabhorn Research Institute under the presidency of Professor Dr.HRH. Princess Chulabhorn has taken an active role in using science and technology to improve the quality of life for the people. Within this concept, one of the Institute's programs was designed to provide a forum for scientists, government officials, industrialists to freely communicate and exchange ideas. This will reduce the communication gap between people working in different sectors. As a part of this program, BESMA international conference was organized under the leadership of HRH. Princess Chulabhorn with the support of United Nation Development Program to provide scientific excellence and also to bridge the technology gap between developed and developing countries.

In the past decade, major breakthroughs in biotechnology and environmental science were achieved as a result of basic research in the areas of microbiology, genetics and biochemistry. This reflects a multi-disciplinary nature of current research and development which makes it necessary for scientists to effectively communicate, exchange ideas and results. To facilitate this process, the conference was organized into four major research areas namely eukaryotic, microbial, plant and environment. This permitted scientists, industrialists and officials working in different disciplines to participate and learn about other research areas not closely related to their own. In this regard, BESMA was an unique conference in that it was successful in bring top rated scientists from highly diverse fields to participate under one conference. Thus, not only theparticipants but also the speakers benefited from the wide range of topics offered.

The conference brought together over 400 participants from 19 different countries from four continents. The key note address on recombination in mammalian cells was delivered by Professor P.Berg, a noble laureate. Altogether, there were eighteen plenary lectures and thirty short oral presentations in four different fields. Over fifty high quality posters were presented. This book contains a collection of papers by invited speakers, reflecting major advances in the areas of eukaryotic, microbial, plant, environment biotechnologies and new technologies. It is also containing topics which have not previously been complied in a single volume.

The Editors

ACKNOWLEDGMENTS

The organizing committee wish to thank all invited speakers who have participated in the conference and contributed to this proceeding. Mahidol university has provided personnel support through out the conference. United Nation Development Program(UNDP) has provided the financial support for the conference. The following organizations also made financial contributions, UNESCO, Thai Airway International Airline, Science Technology Development Board(STDB) and the Rockefeller Foundation. Mr. Suchat Udomsopagit has been most helpful before and after the conference.

CONTENTS

SECTION I
EUKARYOTIC BIOTECHNOLOGY

SECTION II
AGRICULTURAL BIOTECHNOLOGY

SECTION III
ENVIRONMENTAL BIOTECHNOLOGY

SECTION IV
MICROBIAL BIOTECHNOLOGY

SECTION V
NEW TECHNOLOGY

MODIFICATION OF SPECIFIC CHROMOSOMAL LOCI IN MAMMALIAN CELLS BY RECOMBINATION

Paul Berg

Beckman Center, Department of Biochemistry
Stanford University School of Medicine, Stanford,
California 94305

Biochemistry and genetics epitomize what is called the reductionist approach to biology; single biological elements (for example, a protein, an enzyme, or a gene) are examined in great detail through a series of narrowly defined questions and experimental protocols. This approach can illuminate many phenotypic characteristics of complex systems, including whole organisms, if those phenotypes are the direct consequence of a single gene's function.

The recombinant DNA technology has made essentially every segment and virtually any gene from any genome accessible for molecular analysis. But the nucleotide sequence of a gene and its flanking segments alone do not tell us how the gene works or how its expression is regulated during development and differentiation or in response to environmental changes. Neither does the sequence reveal how gene expression is coordinated to ensure the intricate physiological balance characteristic of healthy cells and organisms. Nor can we deduce how the production of a faulty gene product or an altered rate of gene expression will disrupt normal function or produce disease. To comprehend the physiological significance of the molecular details requires biological analysis. Here, too, recombinant DNA techniques provide a powerful experimental approach. When cloned genes are introduced into cells by transfection or injection, they are often expressed and even respond to specific regulatory influences. This provides a way to analyze the relative activities of mutant alleles, and the experimentally constructed mutants can be assessed in this way. Thus, the really special consequence of the recombinant DNA development is that studies of the interdependence of structure and function can now be undertaken *in vivo* as well as *in vitro*. This analytic approach, reverse genetics, replaces many of the methods of classical genetics and expands the potential of genetic research in extraordinary ways.

The transformation of eukaryotic cells in culture by recombinant vectors containing genes and their specifically mutagenized versions is one of the basic techniques in molecular genetics. In such experiments, the newly introduced DNA is incorporated into the host cell's genome at nearly random locations rather than at its original homologous site. This occurs

Biotechnology and Environmental Science: Molecular Approaches
Edited by S. Mongkolsuk et al., Plenum Press, New York, 1992

1

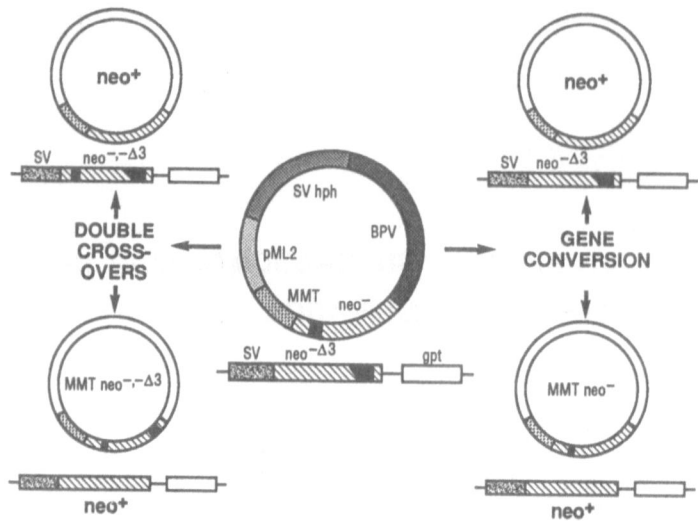

Figure 1. Possible outcomes of homologous recombination between chromosomal and episomal *neo* genes.

Shown in the center of the figure are the participants in the recombination: the bovine papilloma virus (BPV) based episome containing the *neo* gene with the insertion mutation at the 5' end (*neo*) and the chromosomal *neo* gene with the deletion mutation near the 3' end (*neo*[3]). The episomal *neo* allele is transcribed from a mouse metallothionine gene (MMT) promoter while the chromosomal *neo*[3] allele is transcribed by an SV40 (SV) promoter. The outcomes shown to the left and right are alternative ways to account for repair of the chromosomal or episomal *neo* genes. The scheme at the left supposes a reciprocal exchange between homologous parts of the episomal and chromosomal *neo* alleles; the sketch at the right portrays a non-reciprocal exchange or gene conversion. Our data show that all of the recombinations occur by a gene conversion type mechanism.

because most eukaryotic cells are far more proficient at nonhomologous versus homologous recombination. This fact introduces unwanted complexities in the interpretation of such transformations. Thus, the expression and regulation of the introduced gene is often influenced by its integration site. Consequently, the nonhomologous integration of introduced DNA into ectopic locations obscures the regulatory influences and interactions inherent in a gene's normal chromosomal location. Moreover, such nonhomologous integrations may interrupt essential genes and produce undesirable recessive or dominant mutations. As a consequence, a great deal of effort has been made to devise procedures for modifying specific chromosomal loci by recombination -- an experimental approach referred to as gene targeting. In this paper, I describe two approaches carried out in my laboratory towards this end. One of these is the work of a graduate student, David Strehlow, which is described in his Ph.D. thesis and will be reported elsewhere shortly. The second approach was pioneered by a colleague, Maria Jasin (1) and exploited subsequently in collaboration with S.F. Elledge and R.W. Davis (2). Several general review articles on this topic are available (3-5).

One line of experiments aimed to repair a mutation in a chromosomal gene by recombination with a suitably modified version of that gene carried on an autonomously replicating episome (Figure 1). The episomal DNA could be maintained as an autonomously replicating element at about 25-50 copies per cell because it contained the bovine papilloma virus replicating machinery which normally ensures that virus' episomal state in infected cells. Furthermore, the episomal plasmid contained a gene which confers hygromycin resistance (*hph*); therefore, growth of cells in hygromycin ensured continued maintenance of the episome. The chromosomal target was a neomycin resistance gene (*neo*) that had been linked to mammalian transcription signals (SV) and introduced into cultured mouse cells; a deletion of about 180 bp at the 3' end of the *neo* gene's coding region (*neo*³), marked with a XhoI linker rendered the gene inactive. The episomal form of the some *neo* gene transcribed from the metallothionine gene promoter (MMT) carried an inactivating insertion of about 70 bp near the 5' end of the coding region (*neo*) . In our experiments, the plasmid containing the *neo* gene was introduced into cells containing about 10 copies of the chromosomal *neo* ³ gene, and maintained by selection in hygromycin. Note that the episomal *neo* gene contains the sequence deleted from the chromosomal *neo*³ gene and vice versa.

Under these circumstances, cells carrying both mutated *neo* genes segregate *neo⁺* offspring with an unexpectedly high frequency: about 10^{-4}. Examinations of the structures of the chromosomal and episomal *neo* alleles revealed that about half of the *neo⁺* segregants had a completely repaired and functional *neo* gene on the plasmid while the *neo* sequences remained unchanged in the chromosome. The remaining *neo⁺* segregants have a repaired and functional *neo* gene in the chromosome and unchanged *neo* genes in the plasmids. Thus, either the 70 bp insertion or the 180 bp deletion can be repaired equally efficiently by the recombination partner's corresponding wild type sequence. Moreover, the recombinational event is non-reciprocal, a process referred to as gene conversion. The mechanism of such a non-reciprocal exchange of DNA sequence information between two DNA duplexes is presently unknown but its existence is well documented.

Besides being able to repair mutations, gene conversion can also be used to introduce mutations into a wild type sequence. Thus, in principle, any chromosomal gene can be viewed as a target for mutation by an episomal version of all or part of the gene containing the desired mutation. In effect, the plasmid serves as a sequence specific mutagen, possibly able to create every conceivable type of mutational alteration into the targeted gene. Such episomal plasmids are now being developed for a variety of cell types but particularly ones that can be used to introduce mutations into selected genes of mouse embryonic stem (ES) cells. Such modified cells can then be introduced into early mouse embryos (the blastocyst), whereupon they can populate all the tissues, including the germ line, of the newborn mice (5). In this way, mouse strains carrying specific mutations in selected genes can be obtained for use as animal models of human genetic diseases, or for studies of the mutationally induced physiologic responses.

Our second approach to modifying specific chromosomal loci aims to use mammalian cells' ability to repair gaps in one DNA using another DNA that possesses sequences spanning the gap. A model to explain this process -- double strand gap repair -- was proposed by Szostak et al. (6) and is diagrammed in Figure 2. The principal feature of that model is that a double strand duplex is used as the template for replacing the missing sequence in the gapped duplex. An important consequence of this model is that there are two alternate ways to resolve the recombination intermediate shown as the fourth structure from the top.

Cleavage and ligation of the crossing strands at the two places marked [a] leads to separation of the two partners, each with a fully intact and identical sequence over the region corresponding to the gap. However, an equally probable set of cleavages and ligations can occur at positions labeled [b]. In that case, the two duplexes become joined, that is, the repaired gapped DNA and the originally intact DNA are recombined. Where the gapped DNA is a plasmid DNA and the other partner is its chromosomal homologue, the plasmid DNA is integrated into the homologous site.

Using this approach, we have successfully integrated such gapped plasmids at a unique chromosomal target (1). In our experimental design, the gap repair reconstituted a functional promoter which, because it could transcribe a selectable gene marker, allowed us to detect the cells in which the recombination had occured. The frequency with which this occurred was unexpectedly high ($\sim 10^{-5}$).

We have since used a similar strategy to disrupt the human CD4 gene (2). This gene is expressed on the surface of T-helper cells and is needed for the proper signaling of such cells by antigen. Our strategy here was to introduce a gapped DNA containing homology to the CD4 gene so that homologous integration would both disrupt the CD4 gene and create a modified gene able to express a new surface protein. Such successfully targeted cells were detected with antibodies specific for the new surface protein. Here too the frequency is high enough ($\sim 10^{-6}$) to be useful for making gene disruptions in the embryonic stem cell system.

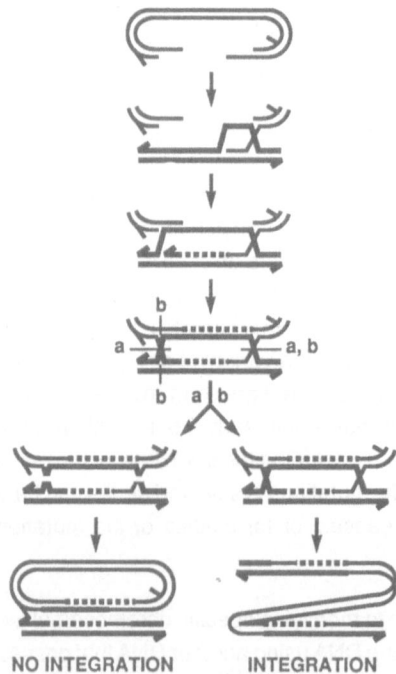

NO INTEGRATION   INTEGRATION

Figure 2. The Szostak et al. (6) model for recombination via double strand gap repair See text for explanation.

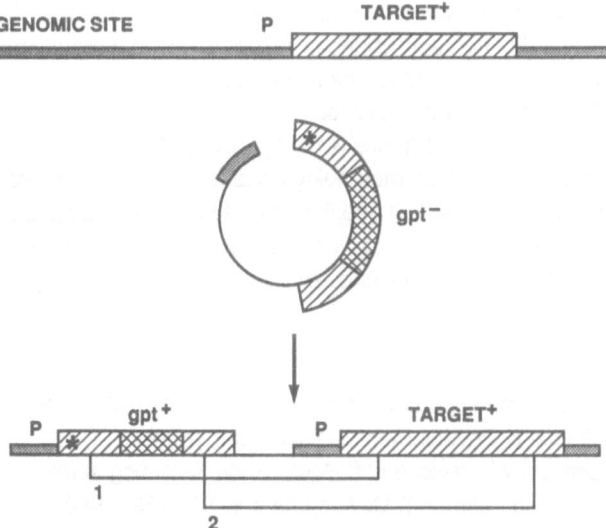

Figure 3. A scheme for chromosomal gene replacement or alteration. See text for explanation.

Recombinational gap repair accompanied by homologous integration may also permit gene replacements as well as the creation of specific alterations into selected genes. The strategy by which this could be accomplished is diagrammed in Figure 3. Consider a target gene at a specific genomic site; P represents the regulatory sequences upstream of the gene's coding region. The integrating DNA contains a gap corresponding to the P sequence; the ends flanking the gap are homologous to the region upstream of P and to the gene's coding sequence. In this case, the coding sequence in the plasmid is interrupted by a selectable gene, *gpt*. In that design, the acquisition of a promoter confers a selectable phenotype on the cells in which the targeted recombination has occurred. The end result of such a homologous insertional integration is a tandem repetition of identical coding sequences as shown at the bottom. The brackets indicate two possibilities for intrachromosomal recombination, the consequences of which are to eliminate the regions contained within each bracket. Note that recombination between homologous sequences in the 3' region (bracket 2) leads to replacement of the target gene by the newly introduced DNA. This would result in a knockout of the target gene on that chromosome. Recombination between sequences near the 5' regions (bracket 1) would lead to reconstitution of the original gene, and loss of the introduced DNA. However, if the introduced DNA carried a mutational alteration (symbolized by the `), recombinational loss at the newly created chromosomal arrangement, might prove useful for introducing any of a variety of mutational alterations into specific genes.

COMMENT

Biologists have long been interested in studying the effects of mutations on the phenotypes of whole cells and organisms in order to understand normal and abnormal states. Before the recombinant DNA breakthrough, spontaneously occurring and induced mutations provided the materials for such analyses. But now biology has changed from its traditional pursuit of understanding how living things are constructed and function to a manipulative science that can make willful heritable changes in cells and whole organisms. Being able to accomplish such alterations accounts for the most profound implications of the recombinant DNA technology. It is here that the reductionism inherent in the analyses of the structure, expression and regulation of single genes is giving way to studies of the pleiotropic effects of single genes on the physiological, morphological and developmental properties of whole cells and organisms. Such information and the ability to engineer specific types of genomic modifications should have important ramifications not only for basic science but also for practical advances in medicine and agriculture.

REFERENCES

1. M.Jasin and P.Berg. Homologous Integration in Mammalian Cells Without Target Gene Selection. Genes & Development 2: 1353-1363 (1988).
2. M.Jasin, S.J. Elledge, R.W. Davis, and P.Berg. Gene Targeting at the Human CD4 Locus by Epitope Addition. Genes & Development 4: 157-166 (1990).
3. S.Subramani and B.L. Seaton. Homologous Recombination in Mitotically Dividing Mammalian Cells. Chapter 18 in Genetic Recombination (R.Kucherlapati and G.R. Smith, Eds.) American Society of Microbiology, Washington, D.C., 1988.
4. R.Kucherlapati and P.D. Moore. Biochemical Aspects of Homologous Recombination in Mammalian Somatic Cells. Chapter 19 in Genetic Recombination (R.Kucherllapati and G.R. Smith, Eds.). American Society of Microbiology, Washington, D.C., 1988.
5. M.R. Capecchi. The New Mouse Genetics: Altering the Genome by Gene Targeting. Trends in Genetics Research 5: 70 (1989).
6. J.W. Szostak, T.L. Orr-Weaver, R.J. Rothstein, and F.W. Stahl. The Double-Strand-Break Repair Model for Recombination. Cell 33: 25 (1983).

MOVEABLE ELEMENTS IN THE HUMAN GENOME: STATUS OF RESEARCH

Maxine F. Singer, Thomas G. Fanning*, Debra M. Leibold,
Gary D. Swergold, and Ronald E. Thayer

Laboratory of Biochemistry, National Cancer Institute, Bethesda,
Maryland 20892 (U.S.A.)

*Present address: Department of Cellular Pathology
Armed Forces Institute of Pathology, Washington, D.C. 20306

The first moveable DNA elements to be studied at the molecular level were the transposable elements found in prokaryotes. Although several different types are now known, they all share two properties: first, the DNA within the element encodes a gene or genes that are required for transposition, and, second, specific DNA sequences are repeated in inverted orientation at the two ends of the element and are required for transposition.

Eukaryotic moveable elements are like the prokaryotic ones; they encode genes required for transposition and have specific terminal DNA sequences. However, two major groups of these elements have been distinguished. One is characterized by relatively short terminal repeated sequences and transposition mechanisms that rely on DNA transactions; the P elements of Drosophila and the activator (Ac) elements of maize fall in this group. The second group, generally termed retrotransposons, utilize transposition mechanisms involving RNA intermediates and reverse transcription; the required reverse transcriptase is encoded within the element. One type of retrotransposon, class I, is characterized by a central segment surrounded by long direct terminal repeats, or LTRs. Class I elements are very similar to retroviruses in structure and, as far as is known, in mechanisms of transposition; they differ from retroviruses in that they lack genes encoding envelope proteins, have no viable extracellular from, and are not known to be infectious. The best characterized elements in this group are the Ty elements of Saccharomyces cerevisiae. Class II retrotransposons have no terminal repeats, and one end often has a stretch rich in A·T base pairs. Although less well understood than class I retrotransposons, the class II elements also occur in a wide variety of eukaryotes.

Regardless of the type of element, insertion of a moveable element into a new genomic locus is generally accompanied by duplication of a short DNA sequence at the target site. These duplications flank the two ends of the insertion. The size of the duplicated DNA sequence varies from one element to another. Elements like P and activator as well as class

Biotechnology and Environmental Science: Molecular Approaches
Edited by S. Mongkolsuk et al., Plenum Press, New York, 1992

7

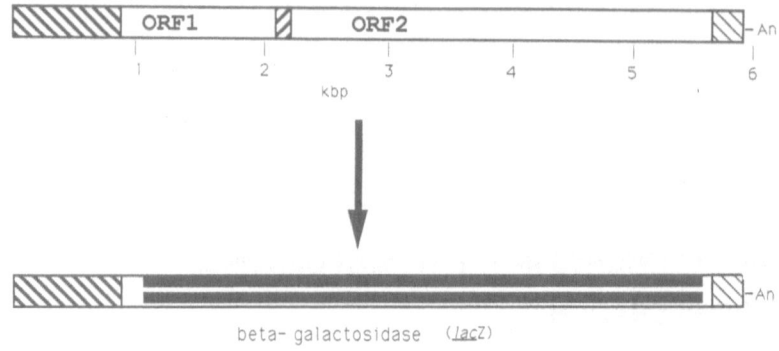

Figure1. Schematic drawing of a full length L1Hs element (top) and of a construct in which the E. coli *lac* Z gene is fused, in frame, after the 15th codon of the L1Hs ORF1 (bottom).

I retrotransposons are generally associated with specific size target site duplications. Class II retrotransposons, however, cause variable size target site duplications.

Two families of dispersed repeated DNA sequences are candidates for moveable elements in the human genome: THE-1 (1) and LINE-1, abbreviated L1Hs (2). The approximately 10^4 THE-1 elements in human DNA contain 2.3 kbp of DNA bounded by 360 bp long LTRs. No coding regions have been identified in isolated THE-1s and nothing is know concerning their capacity to move or the mechanisms that might be involved. Thus, THE-1s remain 'candidate' moveable elements. In contrast, LINE-1 sequences are the most abundant long interspersed repeated elements in mammalian genomes and have all the characteristics of class II retrotransposons. They occur in all mammalian genomes analyzed and in each species they have similar organizations and long stretches of highly homologous coding regions. LINE-1 sequences have no terminal repeats, are typically between 5 and 7 kbp long, are often truncated, and are usually surrounded by variable-length target site duplications. The elements contain one or more open reading frames, often a short 5' ORF1 and a long 3' ORF2, and terminate in a 3' A-rich stretch on the strand containing the ORFS. Figure 1 (top) shows the structure of a full-length (non-truncated) human LINE-1 (L1Hs).

One or more of the approximately 4×10^3 full length L1Hs elements dispersed in the human genome appear to be active transposable elements. For example, two males with hemophilia A have been identified in which the causative mutation is an insertion of an L1Hs element in an exon of the factor VIII gene (3). In each case, both factor VIII genes in the maternal genomes were normal; transposition appears to have occurred during germ cell maturation or very early in embryogenesis. Transposition can also occur in somatic cells. A breast adenocarcinoma has been identified in which an L1Hs element is inserted in one myc allele in the tumor, but not in the surrounding normal tissue (4). Because the second myc allele in the tumor was also aberrant it is not known whether the L1Hs insertion had any influence on oncogenesis.

The approximately 10^5 truncated L1Hs copies in the human genome appear to be nonfunctional; they generally lack sequences starting at the 5' end of full length elements and some contain fewer than 100 L1Hs bp from the far 3' end. Of the full length elements, at least some are transcribed and the RNA are transported to the cytoplasm where they appear as full-length, unspliced, polyadenylated RNAs; the 5' end of the RNAs coincides with residue number 1 at the 5' end of the full-length elements (5). However, these cytoplasmic RNAs were only detectable in human teratocarcinoma cells, NTera2D1 (6). Evidence for less abundant transcription in JEG-3 choriocarcinoma cells was also obtained (5). Thus, efficient transcription appears to be restricted to certain cell types.

cDNA clones prepared from the full-length cytoplasmic L1Hs RNA in NTera2D1 cells gave a consensus sequence in which ORF1 was preceded by an approximately 900 bp long 5' leader with only one or two short open reading frames initiated by AUG codons. ORF1 (1014 bp) and ORF2 (3825 bp) are in the same frame and are separated by 33 base pairs bracketed by two conserved, in-frame translational stop codons. ORF2 is followed by an approximately 200 bp long noncoding, 3' trailer. A consensus L1Hs sequence derived from genomic clones yields the same overall organization although the cDNA and genomic consensus sequences differ from one another (5,7).

Analysis of 19 cDNA clones revealed that no two were identical. This suggests that a large number of the full length L1Hs genomic elements can be transcribed. Two details of these experiments are of special interest. First, either ORF1 or ORF2 or both, in most or all of these cDNA clones, has one or more base pair changes that results in stop codons, thereby making the RNAs incapable of synthesizing either the ORF1 or ORF2 product or both. Several of the cDNAs have open ORF1s. The analysis is incomplete and it remains possible that both ORFs are open in one or more of the cDNA clones. Second, the fact that so many different L1Hs elements are transcribed into full-length RNAs specifically in NTera2D1 cells suggests that all these elements, presumably dispersed, are under the same transcriptional control. However, there is no evidence to suggest that different L1Hs elements have similar flanking sequences. Therefore, it has been suggested that the cis acting sequences required for transcriptional control may exist within the element. This would be consistent with the fact that the promotor for the Drosophila class II retrotransposon jockey has been shown to be within the 5' untranslated region of the element (8).

In order to test whether transcriptional control regions occur within L1Hs sequences, G. Swergold (15) fused the E. coli beta-galactosidase gene, in frame, after the first 15 codons of an L1Hs ORF1 derived from a cDNA clone (Figure 1, bottom). The 5' leader and 3' trailer sequences were typical L1Hs sequences obtained from cloned elements or from synthetic oligonucleotides. This construct promotes the expression of active beta-galactosidase in various transiently transfected mammalian cells. Deletion of the bulk of the 5' leader region virtually abolishes expression. In NTera2D1 cells, the complete construction was approximately 75 percent as active as the SV40 early promoter driving the same reporter gene. However, in HeLa cells, the construct gave only 1 percent as much enzyme as the SV40 promoter. Northern blotting experiments with lacZ probes confirmed that the effect was on transcription. Swergold concluded from these experiments that transcriptional regulatory sequences do occur within the L1Hs 5' leader sequence and moreover, these sequences confer the cell-type specificity for teratocarcinoma cells. The specific cis acting DNA sequences in the 5' leader are being investigated by analysis of deletion derivatives of the construct at the bottom of Figure 1. These experiments have demonstrated that the most critical sequences for transcription are within the first 100 base pairs from the 5' end. However, sequences spread throughout the 5' leader region are also important for full expression.

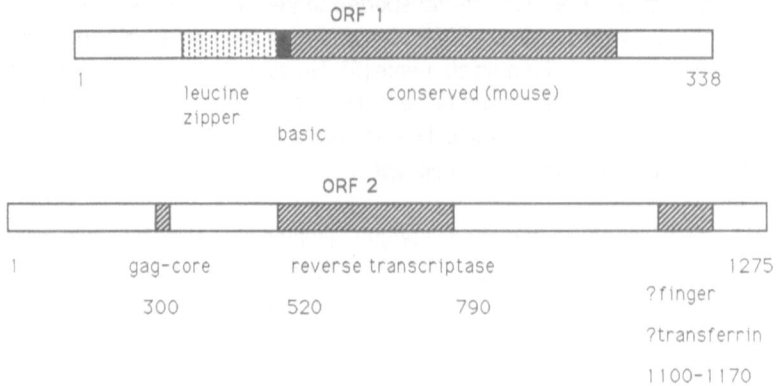

Figure 2. Schematic drawing showing features deduced from the predicted amino acid sequence of L1Hs ORF1 (top) and ORF2 (bottom).

Models for the transposition mechanism for L1Hs propose, as a first step, the synthesis of a full length, L1Hs RNA. This RNA is presumed to serve as a messenger RNA for the synthesis of L1Hs ORF1 and ORF2 proteins. The experiments just described are consistent with this model. The models further propose that the same RNA serves as a template for RNA-dependent DNA synthesis, catalyzed by reverse transcriptase; this DNA would then be inserted in a new genomic position. it appears likely that fully active L1Hs elements encode reverse transcriptase. An approximately 750 bp region within ORF2 encodes a polypeptide with marked homology to known reverse transcriptases (9,10) (Figure 2, bottom); additional regions have homology to the gag-core region of retrovirus proteins (10) and to a zinc finger type structure (9) and transferrin (11). The significance of these observations remains to be demonstrated as no ORF2 proteins have been identified and no such activities detected.

A 38 kDa polypeptide likely to be encoded by ORF1 can, however, be detected in NTera2D1 cells and in another human teratocarcinoma cell line, 2102Ep, as well as in JEG-3 cells (12). A protein which may be the same is barely detectable in Hela and 293 (embryonal kidney) cells. These experiments (12) utilized an antiserum prepared against a fusion protein synthesized in E. coli from a plasmid containing ORF1 fused to the E. coli trpE gene. Both immunocytochemical experiments and cell fractionation experiments indicate that the immunoreactive polypeptide is more abundant in cytoplasm compared to nucleus. The cell-type distribution of the polypeptide is consistent with the previously described experiments indicating that expression of L1Hs is more efficient in certain cell types (e.g., teratocarcinoma) than in others. No clearly significant homologies were found in the predicted ORF1 polypeptide to any sequences in either gene or protein data banks, except for homology to the mouse LINE-1 sequence (13). However, examination of the predicted amino acid sequence did reveal the presence of a "leucine zipper" (14), followed by a basic region (Figure 2,top).

All the experiments carried out thus far are consistent with the models that have been proposed for L1Hs transposition and for its classification as a class II retrotransposon. The fact that L1Hs transposition is a significant mutagenic event in humans has stimulated our interest to investigate further the many unknown aspects of the L1Hs transposition mechanism.

ACKNOWLEDGEMENTS

We thank Beth A. Dombroski and Haig H. Kazazian, Jr. of The Johns Hopkins University for their advice, collaboration and materials. We are also grateful to Skorn Mongkolsuk of the Chulabhhorn Research Institute who carried out preliminary experiments while he was a postdoctoral fellow in our laboratory.

REFERENCES

1. K.E. Paulson, A.G. Matera, N. Deka, and C.W. Schmid. 1987. Nucleic Acids Res. 15 5199-5215.
2. C.A. Hutchison III, S.C. Hardies, D.D. Loeb, W.R. Shehee, and M.H. Edgell. 1989. In Mobile DNA, D.E. Berg and M.M Howe, eds. American Society for Microbiology, Washington, D.C. pp. 593-617.
3. H.H. Kazazian Jr., C. Wong, H. Youssoufian, A.F. Scott, D.G. Phillips, and S.E. Antonarakis. 1980. Nature 332, 164-166.
4. B. Morse, P.G. Rothberg, V.J. South, J.M. Spandorfer, and S.M. Astrin. 1988. Nature 333 87-90.
5. J. Skowronski, T.G. Fanning, and M.F. Singer. 1988. Mol. Cell. Biol. 8 1385-1397.
6. J. Skowronski and M.F. Singer. 1985. Proc. Natl. Acad. Sci USA 82 6050-6054.
7. A.F. Scott, B. J. Schmeckpeper, M. Abdelrazik, C.T. Comey, B. O'Hara, J.P. Rossiter, T. Cooley, P. Heath, K.D. Smith, and L. Margolellt. 1987. Genomics 1 113-125.
8. L.J. Mizrokhi, S.G. Georgieva, and Y.V. Ilyin. 1988. Cell 54 685-691.
9. T. Fanning and M. Singer. 1987. Nucleic Acids Res. 15 2251-2260.
10. R.F. Doolittle, D-F. Feng, M.S. Johnson, and M.A. McLure. 1989. Quart. Rev. Biol. 64 1-30.
11. M. Hattori, S. Hidaka, and Y. Sakaki. 1985. nucleic Acid Res. 13 7813-7827.
12. D.M. Leibold, G.D. Swergold, M.F. Singer, R.E. Thayer, B.A. Dombroski, and T.G. Fanning. 1990. Proc. Natl. Acad. Sci. USA. 87 6990-6994.
13. W.R. Shehee, S-F. Chao, D.D. Loeb, M.B. Comer, C.A. Hutchison III, and M.H. Edgell. 1987. J. Mol Biol. 196 757-767.
14. M. Johnson and S.L. McKnight. 1989. Annu. Rev. Biochem. 58 799-839.
15. G.D. Swergold. 1990. Mol. Cell. Biol. 10 6718-6729.

HOST-VECTOR SYSTEM AND REVERSE GENETICS IN
A NON-CONVENTIONAL YEAST, Candida maltosa

Masamichi Takagi

Department of Agricultural Chemistry, The University of Tokyo
Bunkyo-ku, Tokyo 113, Japan

In Candida maltosa, expression of many genes is induced by the hydrophobic compound, n-alkane, resulting in biogenesis of organelles such as endoplasmic reticulum and peroxisome where n-alkane is metabolized. Host-vector systems are needed in order to investigate this phenomenon at the molecular level. Since C. maltosa does not have plasmids, the isolation of an ARS site from the genome of C. maltosa was attempted so as to construct vectors. Several host-vector systems were constructed to promote reverse genetics in this non-conventional yeast as described in our recent papers shown below.

1) Construction of a host-vector system in Candida maltosa by using an ARS site isolated from its genome (J. Bacteriol., 167, 551-555, 1986):

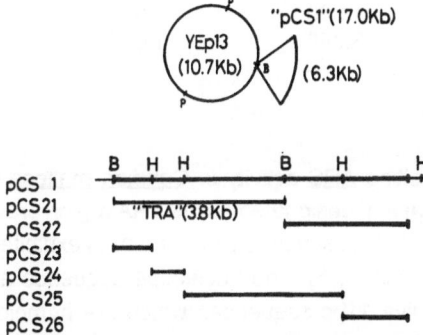

Fig. 1 Subcloning of an *ARS* site of *C. maltosa*. (Top) pCS1 (17.0 kb) is the original plasmid isolated from one of the Leu+ transformants obtained by transformation of *C. maltosa* J288 (*leu*) with a gene library of the *C. maltosa* genome prepared in the YEp13 (*LEU2*+) vector. (Bottom) Six kinds of plasmids (pCS21 through pCS26) were constructed in the YEp13 vector, each containing one of the six restriction fragments, and they were transformed into *C. maltosa* J288 (*leu*) to subclone an *ARS* site. Since pCS21 had *ARS* activity (see Table 1), the 3.8-kb fragment was designated the TRA region. Heavy lines represent the *C. maltosa* inserts. Abbreviations: P, *Pst*I; B, *Bam*HI; H, *Hind*III.

Biotechnology and Environmental Science: Molecular Approaches
Edited by S. Mongkolsuk et al., Plenum Press, New York, 1992

13

To construct a host-vector system in an n-alkane- assimilating yeast, Candida maltosa, the isolation of an ARS site from its genome which replicates autonomously in C. maltosa was attempted. Leu⁻ mutants of C. maltosa were transformed with a gene library prepared by using YEp13 (LEU2⁺) as a vector, and Leu⁺ transformants were obtained at a high frequency. A plasmid named pCS1 was isolated from the recipient cells. pCS1 contained a 6.3kb fragment of the C. maltosa genome, and a 3.8kb fragment with ARS activity was subcloned and designated the TRA (transformation activity) region (Fig. 1, 2). Vectors (pTRA1 and pTRA11) for C. maltosa J288 were constructed that contained this 3.8kb fragment, pBR322, and LEU2 gene of Saccharomyces cerevisiae (Fig. 3). Transformation of C. maltosa J288 with these plasmids was successful by both spheroplast and lithium acetate methods (Table 1). Southern blot analysis suggested that the copy number of pTRA1 in C. maltosa was between 10 and 20, and it was stably maintained during growth without selective pressure in the medium. It was also found that these vectors could transform S. cerevisiae leu2⁻ to LEU2⁺ , suggesting that the TRA region contained an ARS site (s) that was specific not only for C. maltosa but also for S. cerevisiae.

2) Subcloning and nucleotide sequencing of an ARS site of Candida maltosa which also functions in Saccharomyces cerevisiae (Agric. Biol. Chem., 51, 1587-1591):

Table 1. Frequency of transformations of *C. maltosa* with various plasmids.

Plasmid	No. of transformants/mg of DNA		
	Spheroplast method		Lithium acetate method
	S. cerevisiae	C. maltosa	C. maltosa
YEp13	820	0	0
pBR-LEU	—ᵃ	0	0
pBR-TRA	0	0	0
pARS1	4,180	3	0
pARS11	2,260	0	0
pCS1	1,650	330	280
pCS21	1,320	370	60
pTRA1	3,080	1,650	1438
pTRA11	-	-	439

ᵃ—, Not determined.

A DNA region exhibiting ARS activity in Candida maltosa was subcloned from an isolated 3.8kb DNA fragment and designated as the TRA region previously. it was found that a DNA fragment of about 200bp (designated as fragment 2) exhibited ARS activity in both C. maltosa and S. cerevisiae (Fig. 4, 5), and nucleotide sequence analysis of this fragment revealed that it contained five 11bp-sequences which are homologous to the consensus sequence of ARS sites of S. cerevisiae (9 among 11bp being identical in each, Fig. 6). Plasmids constructed using fragment 2 (pTRA2 and pTRA12) will be useful as cloning vectors, although they showed lower transforming frequencies and lower stabilities (Fig. 7) in C. maltosa than plasmids containing the TRA region (pTRA1 and pTRA11).

3) An improved host-vector system for Candida maltosa using a gene isolated from its genome that complements the his5 mutation of Saccharomyces cerevisiae (Curr. Genet., 16, 261-266, 1989):

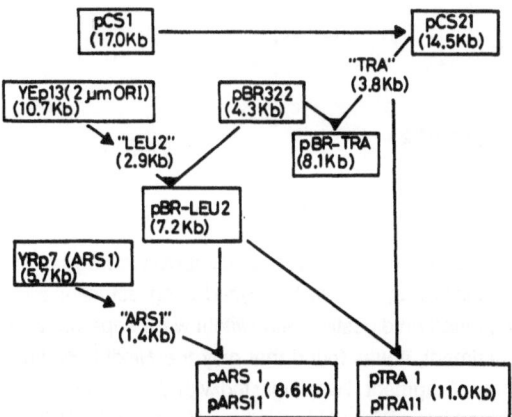

Fig. 2 Various plasmids were constructed to analyze the function of the TRA region. The names of the plasmids used in the transformation experiments summarized in Table 1 are shown in the boxes. The orientation of the TRA region into pBR-LEU was different in pTRA1 and pTRA11, and that of the *ARS1* site was different in pARS1 and pARS11. The approximate sizes of the DNA fragments and plasmids are shown in parentheses.

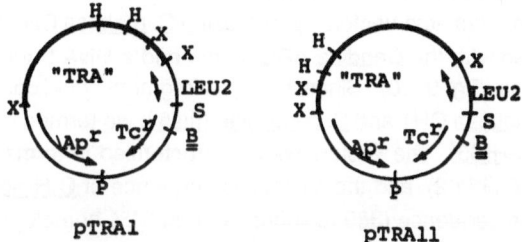

pTRA1   pTRA11

Fig. 3 Structure of pTRA1 and pTRA11. These two plasmids are 11.0 kb, and the *Bam*HI site is the convenient site. The TRA region (3.8 kb, heavy lines) came from the genome of *C. maltosa*. Abbreviations: B, *Bam*HI; S, *Sal*I; X, *Xho*I; H, *Hind*III; P, *Pst*I.

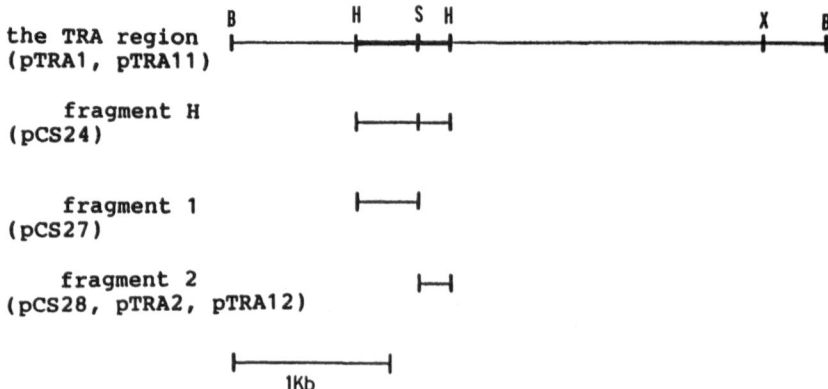

Fig. 4 Restriction maps of the TRA region (3.8 kb) and subcloned fragments of it. The TRA region, cloned As a BamHI-BamHI fragment, was analyzed using restriction enzymes, HindIII and XhoI. The 3 fragments obtained on digestion with HindIII were separately cloned into YEp13. In a transformation experiment, it was found that only the HindIII-HindIII fragment (fragment H) exhibited ARS activity. Fragment H was digested with Sau3AI, and the resulting two fragments were separately cloned into pBR-LEU to produce pCS27 and pCS28 (Fig. 2). Only fragment 2, i.e., not fragment 1, exhibited ARS activity, which is about 200 bp in size. The names of plasmids constructed using each of fragments are given in parentheses. Abbreviations: B, BamHI; H, HindIII; S, Sau3AI; X, XhoI.

The host-vector system of Candida maltosa which we previously constructed utilizes C. maltosa J288 (leu2⁻) as a host. As this host had a serious growth defect on n-alkane (Fig. 8), we developed an improved host-vectors system using C. maltosa CH1 (his⁻) as host. The vectors were constructed with the Candida ARS region and a DNA fragment isolated from the genome of C. maltosa (Fig. 9, 10). Since this DNA fragment could complement histidine auxotrophy of both C. maltosa CH1 and S. cerevisiae (his5⁻), we termed the gene contained in this DNA fragment C-HIS5. The vectors were characterized in terms of transformation frequency and stability (Table 2), and the nucleotide sequence of C-HIS5 was determined. The deduced amino acid sequence (389 residues) shared 51% homology with that of HIS5 of S. cerevisiae (384 residues).

4) Isolation and sequencing of a gene, C-ADE1, and its use for a host-vector system in Candida maltosa with two genetic markers (Agri. Biol. Chem. 55, 59-65, 1991). As the wild-type strain Candida maltosa IAM12247, which is the parental strain of j288 (leu2⁻) and CH1 (his5⁻) described above, is diploid or aneuploid (J.Gen. Appl. Microbiol., 30, 489, 1984), it is expected that most of genes have two alleles. So, two genetic markers in a strain are required to investigate the function of a gene by the gene-disruption technique. Since all of the hosts we have constructed so far have a single genetic marker, we have developed a new host-vector system using a host C. maltosa CHA1 (his5, ade1) with two genetic markers as follows. By UV irradiation of CH1 (his⁻), CHA1 (his5⁻, ade1⁻) was isolated as a mutant forming a red colony and found to be adenine auxotrophy. A DNA fragment harboring complementing

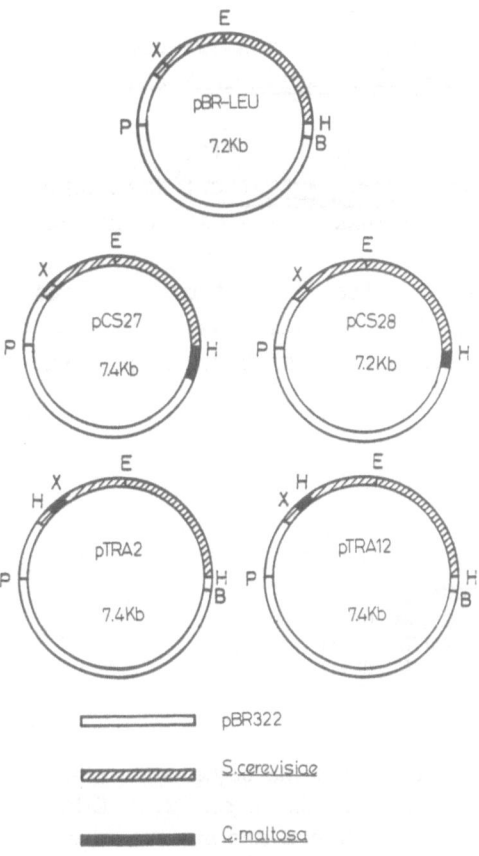

Fig. 5 Plasmids construction. To determine which of the two fragments (fragment 1 and 2 in Fig. 1) exhibited ARS activity, each of them was inserted into the *Bam*HI-*Hin*dIII site of pBR-LEU (7.2 kb) which consists of pBR322 and the LEU gene of *S. cerevisiae* to produce pCS27 (7.4 kb) and pCS28 (7.2 kb), respectively. As only pCS28 exhibited ARS activity, fragment 2 was inserted into the *Xho*I site of pBR-LEU to produce cloning vectors, pTRA2 and pTRA12 (both 7.4 kb). Abbreviations: B, *Bam*HI; E, *Eco*RI; H, *Hin*dIII; P, *Pst*I; X, *Xho*I.

activity of this deficiency was isolated from the C. maltosa genome. Since the DNA fragment also complemented adenine auxotrophy of S. cerevisiae (ade1), we termed a gene contained in this DNA fragment C-ADE1. The deduced amino acid sequence of the C-ADE1 gene product has 65.6% to that of ADE1 (S. cerevisiae) gene product. This homology value is considerably lower than that between the C-LEU2 (C. maltosa) and LEU2 (S. cerevisiae) gene product. This homology value is considerably lower than that between the C-LEU2 (C. maltosa) and LEU2 (S. cerevisiae) gene products, 76% (Curr. Genet., 11, 451, 1987), and is higher than that between the C-HIS5 (C. maltosa) and HIS5 (S. cerevisiae) gene products, 51.1%. Isolation of CHA1, and cloning of C-HIS5 and C-ADE1 make it possible to disrupt two alleles of each gene of C. maltosa. Fig. 11 exhibits the strategy for a two-step gene disruption. Now, we are ready to destroy some gene whose functions are interesting, for example, RIM-C gene (J. Gen. Appl. Microbiol., 31, 267, 1985; J. Bacteriol., 168, 417, 1986) and cytochrome P-450alk (Agric. Biol. Chem., 53, 2217, 1989) to know their functions and the mechanism of their regulation of expression.

```
        10          20          30          40          50          60
GATCAGTTGT ATTTAAGTTG ACAATTTGGT TCACTAAATT TAATTAGTTA GTTGTAAGTC
CTAGTCAACA TAAATTCAAC TGTTAAACCA AGTGATTTAA ATTAATCAAT CAACATTCAG

        70          80          90         100         110         120
AGTTTTGAAA ATTTTAGTCA AATTTTATTA CCAATTTTTT ATGTACACTT AGCAAAAATT
TCAAAACTTT TAAAATCAGT TTAAAATAAT GGTTAAAAAA TACATGTGAA TCGTTTTTAA

       130         140         150         160         170         180
GAAAGCACTT TTATACAGTA GTTACAACGA GCATACGTGG TAGTAATGGA TAATCTATGT
CTTTCGTGAA AATATGTCAT CAATGTTGCT CGTATGCACC ATCATTACCT ATTAGATACA

       190         200
GATATATGAT AAGCTT
CTATATACTA TTCGAA
```

```
C. maltosa        5'           3'
              1  AGTTGTATTTA   (9/11)
              2  ATTTAAGTTGA   (9/11)
              3  AATTAAATTTA   (9/11)
              4  TTTTTATGTA    (9/11)
              5  TTTTATGTACA   (9/11)
Consensus        A        G   A
(S.cerevisiae)   TTTTAT TTT T
```

Fig. 6 Nucleotide sequencing of fragment 2. Fragment 2 was sequenced and the sequence is shown from the *Sau*3AI site to the *Hind*III site (see Fig. 1). DNA sequences homologous to the consensus of ARSs of *S. cerevisiae* (more than 80% homology) were searched for using a computer, and five sequences were found, which are underlined. Arrowheads indicate the direction of the detected sequence. The 5 homologous sequences are summarized in the lower part, together with the consensus sequences of ARSs of *S. cerevisiae*.

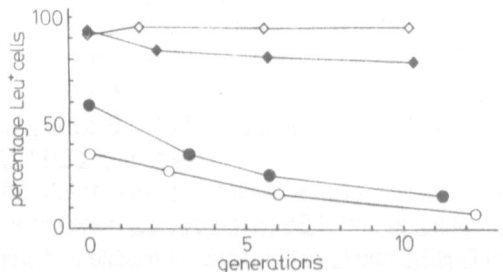

Fig. 7 Stability of pTRA2 and pTRA12 in comparison to that of pTRA1 and pTRA11 in *C. maltosa* grown in a non-selective medium. Cells of *C. maltosa* carrying one of the four plasmids, pTRA1 ( ), pTRA11 ( ), pTRA2 (•) or pTRA12 (○), were grown in a non-selective medium. An aliquot was removed at intervals and plated onto plates to determine the total cell number (generations) and the percentage of plasmid-carrying Leu+ cells. The averages for two independent experiments are presented.

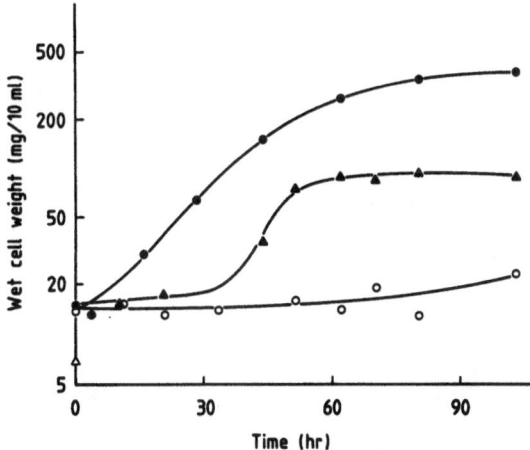

Fig. 8 Growth curves of three kinds of *C. maltosa* in MS medium. Each strain was grown overnight in MY medium and then transferred at an inoculation size of 0.5% into MS medium containing 1% *n*-tetradecane. After cultivation for a period of time, an aliquot was removed, centrifuged and washed once with water, and the wet cell weight was determined. •—• *C. maltosa* IAM12247; ○—○ *C. maltosa* J288 (*Leu⁻*); ▲—▲ *C. maltosa* CH1 (*His-*).

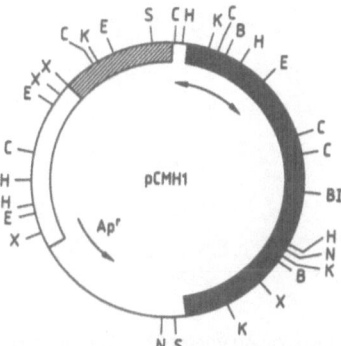

Fig. 9 Physical map of pCMH1 (20 kbp) containing a DNA fragment of *C. maltosa* genome that complements the *His⁻* mutation of *C. maltosa* CH1(*His⁻*). A gene library of *c. maltosa* DNA prepared by using pTRA11 as a vector was introduced into *C. maltosa* CH1. Plasmid DNA of pCMH1 isolated from cells of a *His⁺* colony was analyzed for its structure with various restriction enzymes.   *C. maltosa* genomic DNA;   the *LEU2* gene of *S. cerevisiae*;   the TRA region (3.8 kbp); —the pBR322 sequence; ↔ a EcoRI-HindIII fragment subcloned in the following experiment. *Abbreviations: B* BamHI; *Bl* BglII; *C* ClaI; *E* EcoRI; *H* HindIII; *K* KpnI; *N* NruI; *S* SalI; *X* XhoI .

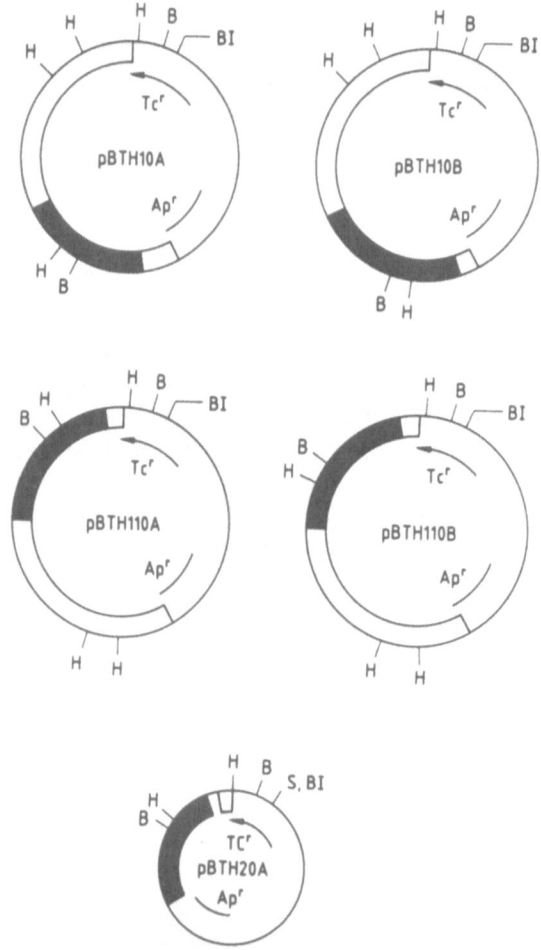

Fig.10 Structure of the cloning vectors for *C. maltosa* constructed by using the *C-HIS5* gene. Four cloning vectors (pBTH10A, pBTH10B, pBTH110A and pBTH110B, each 10.5-kbp) consisted of the *C-HIS5* gene, the pBR322 sequence and the 3.8-kbp TRA region; plasmid pBTH20A (6.9-kbp) contained the 0.2-kbp TRA region instead of the 3.8-kbp fragment. Symbols and abbreviations are the same as described in the legend to Fig. 2.

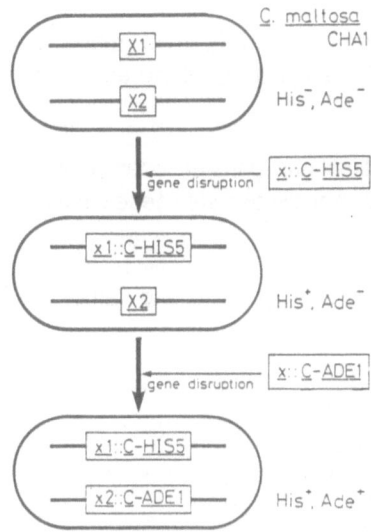

Fig.11 The strategy for a two-step gene disruption in *C. maltosa* strain CHA1. The first step is disruption of one allele of gene with one genetic marker (*C-HI55*). The second step is to destroy the other allele with another genetic marker (*C-ADE1*).

Table 2. Transformation frequency and stability of various vector plasmids.

Plasmid	No. of transformants/mg DNA	Stability (%)
pBTH10A (3.8)[a]	84 (2)[b]	51[c]
pBTH10B (3.8)	247 (4)	88
pBTH110A (3.8)	87 (2)	88
pBTH110B (3.8)	34 (2)	71
pBTH20A (0.2)	90 (2)	0.2

[a] The value enclosed in parentheses is the length (kbp) of the TRA region in the plasmid.
[b] The number enclosed in parentheses indicates the number of experiments.
[c] Percentage stability was calculated by dividing the number of colonies on the histidine-minus plate by the number of colonies on the histidine-plus plate after 63 h of cultivation in the MY medium (average of duplicate experiments).

Some other related papers:

5)   Prification of cytochrome P-450alk from n-alkane-grown cells of Candida maltosa, and cloning and nucleotide sequencing of the encoding gene (Agri. Biol. Chem., 53, 2217-2226, 1989)

6)   Evidence that more than one gene encodes n-alkane -inducible cytochrome P-450s in Candida maltosa, revealed by the two-step gene disruption (Agri. Biol. Chem., 55, 1757-1764, 1991)

7)   A cytochrome P450alk [CYP52] multigene family in Candida maltosa (DNA and Cell Biology, 10, in press, 1991)

EXPRESSION OF THE β-1,3-GLUCANASE GENE IN YEAST *HANSENULA POLYMORPHA*

Shi-Hsiang Shen and Lison Bastien

Biotechnology Research Institute, National Research Council of Canada
6100 Royalmount Ave., Montreal, Quebec, Canada, H4P 2R2

The yeast cell wall is composed mainly of glucan, mannoprotein and chitin (1). In most yeasts, the polysaccharide glucan is predominantly β-1,3-linked with some branching via β-1,6-linkages (2). The glucan network in the yeast cell wall, due to its rigidity, is an essential structure in preventing lysis of the protoplast in a hypotonic environment. Several microorganisms have been reported to produce extracellular enzymes capable of lysing yeast cells (3,4). Analysis of the constituents of these lytic enzyme preparations revealed the presence, among other activities, of β-1,3-glucanases (3,5). When combined with a thiol reagent, β-1,3-glucanase alone was found to be able to lyse the yeast cell (6). Currently, the most common-used enzyme for digestion of yeast cell walls is an enzyme preparation called Zymolyase, produced from *Oerskovia xanthineolytica* (5). We have isolated the glucanase gene from this organism and completely determined its sequence (7).

Since the methylotrophic yeast *Hansenula polymorpha* has been successfully developed as a host for the expression of heterologous proteins (8), we have been interested in expressing this bacterial enzyme in yeast. More over, it would be useful to determine whether

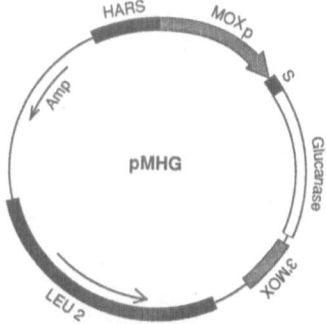

Figure 1. Yeast expression vector. S represents sequence encoding for the signal peptide within the glucanase.

Biotechnology and Environmental Science: Molecular Approaches
Edited by S. Mongkolsuk et al., Plenum Press, New York, 1992

23

the functional expression of glucanase gene in yeasts could affect the cell growth and the secretory properties of yeasts. This question is based on the fact that the rigidity structure of β-1,3-linked glucan, combined with other components, in the yeast cell wall forms a barrier for secretion of some periplasmic proteins, such as the native octamer of invertase and recombinant hepatitis B virus surface antigen (HBsAg), into the culture medium (8,9).

EXPRESSION OF THE β-1,3-GLUCANASE GENE IN THE YEAST

The glucanase gene (7), including sequence encoding the signal peptide, was cloned into an *H.polymorpha* expression vector (8) which contains the promoter of the methanol oxidase gene (MOXp), the autonomous replicon HARS and the *leu*-2 gene from *Saccharom yces cerevisiae* for complementary selection. The resultant expression vector was designated pMHG (Fig.1). The pMHG plasmid was transformed into a leucine auxotroph, *leu*-1, of *H.polymorpha* (8). Transfromants were induced by adding methanol in the culture as a carbon source. After induction, the glucanase activity was detected from the culture medium by a zone lysis method (7). However, it is unknown whether the bacterial signal peptide within the glucanase precursor would be effective for directing processing and secretion of the enzyme protein in *H. polymorpha*. Thus, we investigated the kinetics of glucanase synthesis and secretion into the culture medium. Cultures were sampled daily and the enzyme activity in the cytoplasmic and in the medium was separately determined by digesting glucan substrates (7). Activity assay showed that over 85% glucanase activity was secreted into the medium, indicating that the bacterial signal peptide functions in *H. polymorpha* as effectively as the yeast leader sequences (10). Figure 2 shows the secreted glucanase activity in the medium as a function of time. Secretion of glucanase synthesized in *H. polymorpha* was observed at the first day of induction. The amount of glucanase secreted was gradually accumulated in the medium during induction as the synthesis of the enzyme in the cells continued. After day 6 in induction, the enzyme activity in the medium reached to more than 2 units per millilitre. The yeast cells expressing glucanase grew normally in term of the growth rate and the viability of the cells.

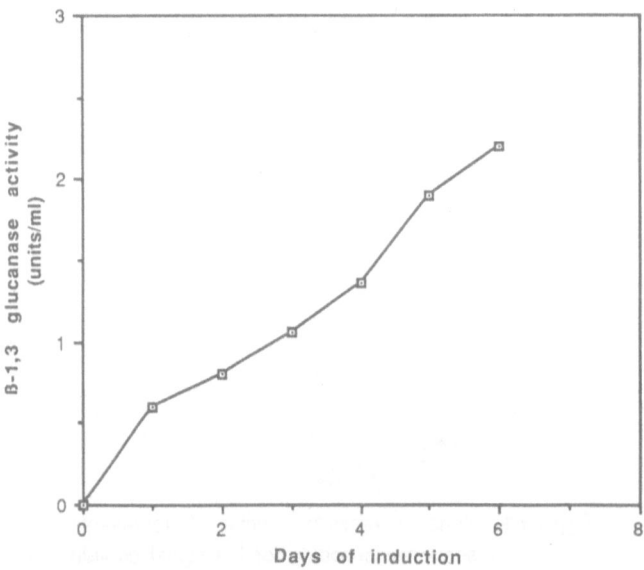

Figure 2. Secretion of glucanase into the yeast medium.

CHARACTERIZATION OF THE EXPRESSED ENZYME

The glucanase enzyme secreted into the medium was characterized by immuno-blotting analysis. As shown in Figure 3, the glucanase protein secreted from yeast immuno-reacted well with the antibodies against the native bacterial enzyme (lane 1). However, the yeast enzyme was a heterogeneous protein. The main reactive component of the yeast glucanase is about a 68 kDa protein. The M_r of glucanase protein without modification in *E. coli*(7) is approximately 57000 (Lane 2). The observed higher molecular mass of yeast enzyme was likely due to the glycosylation since incubation of this yeast enzyme with endo H resulted in reduction of molecular weight of the protein (data not shown). There are seven potential N-linked glycosylation sites (7). It is unclear how many sites, and which sites are glycosylated in this protein.

The yeast enzyme was found to be less active than the bacterial one. It is possible that the heavy glycosylation of the enzyme may have interfered with appropriate folding of the protein, consequently reducing the enzymatic activity.

EFFECT OF YEAST GLUCANASE ON THE SECRETION OF HBsAg FROM YEAST

Our previous experiments showed that the 22 nm particles of HBsAg synthesized in yeast were secreted outside the cell membrane into the periplasmic space, and further excreted into the culture medium following treatment of the yeast cell wall with Zymolyase (8). These results imply that the yeast cell wall might be genetically modified for efficient secretion of proteins by engineering the β-1,3-glucanase gene into yeasts provided that the yeast glucanase is effective to permeabilize the cell wall. To further examine the effect of glucanase activity on the secretion of the HBsAg particles from yeast, the yeast cells expressing HBsAg were incubated with secreted yeast glucanase. As shown in Table 1, while no excretion of HBsAg was observed in the control cells (without glucanase activity), 45% of the antigen was released into the medium by yeast glucanase. The efficacy of yeast glucanase for excretion of HBsAg appeared to be compatible to that of the bacterial enzyme (52 %). Further studies are underway in co-expression of the glucanase gene with the HBsAg gene in yeast cells. We hope that the efficiency of excretion of HBsAg and other complex proteins from yeasts would be greatly improved by co-expression of the glucanase gene. Although more studies are required, our preliminary results seem to be very promising for co-secretion of both HBsAg and glucanase into the medium.

Table 1   Effect of glucanase on the excretion of HBsAg from yeast

	Control	Bacterial glucanase	yeast glucanase
Excretion(%)	0	52	45

In summary, the glucanase produced from yeast, though modified by heavy glycosyla-tion, is still active in digesting glucan substrates and permeabilizing the yeast cell wall for releasing of the periplasmic proteins into the medium. Based on these results, it is expected that a genetically engineered yeast by integration of the glucanase gene into the yeast cells would have a modified cell wall which, while maintaining some rigidity, would possess high porosity that would allow complex supramolecules to be released into the culture medium, in a manner similar to that of animal cells.

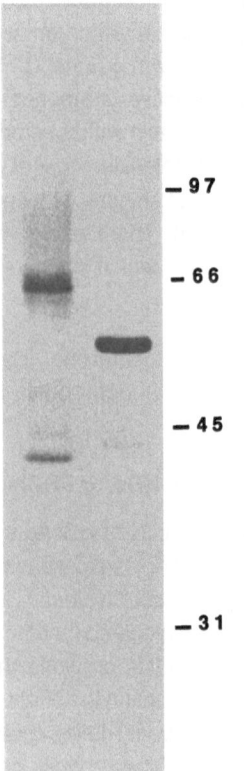

Figure 3. Western blot analysis of gluca-nase. Immunoblotting was carried out as described (7). The numbers are molecular mass maker with sizes given in kDa.

ACKNOWLEDGMENTS

We thank Pierre Chretien for making antibodies and other help in this work.

REFERENCES

1. Ballou, C.E.(1982) in *The Molecular Biology of the Yeast Saccharomyces* (Strathern, J.N., Jones, E.W., and Broach, J.R.,eds)pp. 335-360, Cold Sping Harbor
2. Manners, D.L., Masson, A.J.,and Patterson, J.C.(1973) *Biochem. J.***135**,19-31
3. Doi, K., Doi, A., and Fukui, T. (1971) *J.Biochem.* **70**,711-714
4. Kitamura, K., Kaneko, T., and Yamamoto, Y.(1971) *Arch. Biochem. Biophys.* **145**, 402-404
5. Katamura, K., Kaneko, T., and Yamamoto, Y. (1974)J. *Gen. Appl. Microbiol.* **20**, 323-344
6. Scott, j., and Schekman, R. (1980) *J. Bacteriol.* **142**, 414-423
7. Shen,S.H., Chretien, P., Bastien, L., and Slilaty, S.N. (1991) *J. Biol. Chem.* **266**, 1058-1063
8. Shen, S.H., Bastien, L., Nguyen, T., Fung, M., and Slilaty, S.N. (1990) *Gene* **84**, 303-309
9. Esmon, P.C., Esmon, B.E., Schauer, I.E., Taylor, A., and Schekman, R. (1987) *J. Biol. Chem.* **262**:4387-4394
10. Emr, S.D., Schauer, I., Hansen, N., Esmon, P., and Schekman, R. (1984) *Mol. Cell. Biol.* **4**:2347-2355

ROLE OF THE αA-CRYBP1 SITE IN LENS-SPECIFIC EXPRESSION OF THE αA-CRYSTALLIN GENE

Christina M. Sax*, John F. Klement+, Joram Piatigorsky

Laboratory of Molecular and Developmental Biology, National Eye Institute
National Institutes of Health, Bethesda, MD 20892

INTRODUCTION

A heterogeneous group of crystallins comprise 80-90% of the soluble protein of the ocular lens and are responsible for lens transparency (see 30 for review). The two α-crystallins (αA and αB) are present in all vertebrates, and arose from duplication of a common ancestral gene. The αA gene is expressed in a lens-specific manner, while the αB gene is expressed in numerous tissues (see 24 for review). A combination of transgenic mouse (28), mutagenesis, and cDNA cloning experiments (20) have implicated the αA-CRYBP1 site (5'-GGGAAATCCC-3' at -66 to -57) in the lens-specific expression of the mouse αA-crystallin gene.

The αA-CRYBP1 site bears striking sequence similarity to the consensus sequence for the ubiquitously expressed transcription factors PRDII-BF1 (10), MBP-1 (1), HIV-EP1 (18b), NF-kB (see 17 for review), H2TF1 (2,3), KBF1 (32), EBP1 (8), and HIVEN86 (4). A cDNA which encodes a protein (αA-CRYBP1) that specifically binds to the functionally important -73/-55 region of the mouse αA-crystallin promoter has been isolated from an SV40 T-antigen transformed mouse lens epithelial cell line (20). This cDNA hybridizes to a 10 kb mRNA in lens and other tissues, and codes for a protein with two zinc fingers and 75% sequence identity with the comparable region of human transcription factors PRDII-BF1, MBP-1, and HIV-EP1 (20). Here, we have set out to determine the ability of the αA-CRYBP1 site to confer lens-specific gene expression in light of the apparent ubiquitous distribution of its cognate binding protein.

---

*CORRESPONDING AUTHOR

+present address: Department of Molecular Oncology and Virology, Roche Institute of Molecular Biology, Roche Research Center, Nutley, NJ 07110

Biotechnology and Environmental Science: Molecular Approaches
Edited by S. Mongkolsuk et al., Plenum Press, New York, 1992

27

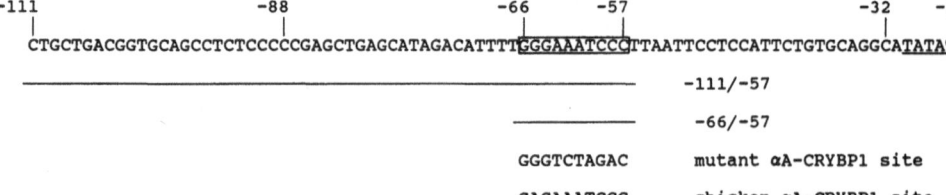

Figure 1. αA-crystallin oligodeoxynucleotides cloned into pTKCAT. The nucleotide sequence of the mouse αA-crystallin -111/-26 region (7) is shown. The αA-CRYBP1 sequence is boxed and the TATA box is underlined. The denoted complementary oligodeoxynucleotides were synthesized with Hind III compatible ends, NICK column (Pharmacia LKB Biotechnology, Inc.) purified, cloned into Hind III-digested pTKCAT, and subsequently used for transfection.

RESULTS AND DISCUSSION

Complementary synthetic oligodeoxynucleotides spanning various regions of the mouse αA-crystallin promoter and containing the αA-CRYBP1 site (Figure 1) were cloned into the plasmid pTKCAT (19) upstream of the Herpes Simplex virus thymidine kinase (tk) promoter fused to the bacterial chloramphenicol acetyltransferase (CAT) gene. The resulting plasmids were co-transfected with pTB1, using calcium phosphate precipitation (14), into an SV40 T-antigen transformed mouse lens epithelial clonal cell line (αTN4-1; 31), a mouse fibroblast cell line (L929; 9), primary embryonic chicken lens epithelial cultures (PLE; 5, 26), and primary embryonic chicken fibroblasts (CEF; 26). pTBI consists of the Rous Sarcoma virus long terminal repeat linked to the bacterial lacZ gene (5). Cultures were harvested 48 hours post-transfection and CAT (21) and β-galactosidase (22) activities were assayed (see Figure 2). Control plasmids pαA366aCAT, which contains the -366/+46 promoter fragment of the mouse αA-crystallin gene fused to the CAT gene, and pSV2CAT and pRSVCAT, were expressed as expected in a lens-specific (6, 20, 23) or ubiquitous (12) manner, respectively (data not shown).

When cloned in the forward orientation at position -130 relative to the tk promoter initiation site, the αA-CRYBP1 site (5'-GGGAAATCCC-3') alone augmented the activity of the tk promoter by approximately 4.7-fold in transfected αTN4-1 cells, but not in L929, PLE, or CEF cells (Figure 2). A similar level of activation in αTN4-1 cells is observed when theαA-CRYBP1 site is cloned in the reverse orientation (data not shown). A mutation (5'-GGGTCTAGAC-3') of the mouse αA-CRYBP1 site did not activate the tk promoter in transfected αTN4-1, L929, PLE, or CEF cell (Figure 2), indicating that the previously observed activation is specific to the αA-CRYBP1 site. This mutation also eliminates transcriptional activity of the αA-crystallin promoter in transfected αTN4-1 cells (20). Although the αA-CRYBP1 site specifies mouse lens-preferred expression in an orientation-independent manner, it appears to differ from classical enhancers in the relatively low level (3-4 fold) of activation and in its requirement for close proximity to the promoter (data not shown).

In contrast to its behavior in αTN4-1 cells, the mouse αA-CRYBP1 sequence alone did not activate the tk promoter in chicken PLE cells (Figure 2). However, inclusion of sequences up to -111 did activate the tk promoter in PLE cells. These results are consistent with a basic difference between αA-crystallin gene regulation in mouse and chicken lenses (7, 16, 27, 28, 18a). The chicken (16) and human (15b) αA-crystallin promoters contain a sequence which differs from the mouse αA-CRYBP1 site by a G to A substitution at position two (5'-GAGAAATCCC-3'). The consensus binding sequence for the related family of NF-kB/H2TF1/

PRDII-BF1/αA-CRYBP1 transcription factors exhibits an absolute conservation of a G residue at this position (17). When the chicken/human αA-CRYBP1 site was cloned in either orientation into pTKCAT it did not increase CAT activity in the transfected αTN4-1, L929, PLE, or CEF cultures (Figure 2). This single base change of the mouse αA-CRYBP1 site is therefore sufficient to eliminate its ability to stimulate the tk promoter in αTN4-1 cells.

Next, oligodeoxynucleotides containing two, three, or four copies of the mouse αA-CRYBP1 site were cloned into pTKCAT upstream of the tk promoter and used for transfection. As in L929, PLE, or CEF cells (Figure 2) a single αA-CRYBP1 site did not activate the tk promoter in transfected S194 cells (15a; Figure 3), a mouse myeloma cell line which constitutively expresses the αA-CRYBP1-related transcription factor NF-kB (29). However, CAT activity increased with copy number of the mouse αA-CRYBP1 site in all transfected cell types (Figure 3). In general, for each plasmid the level of CAT activity produced was as follows: αTN-4 > L929 = S194 > PLE CEF. Transcriptional activation directed by multimers of the mouse αA-CRYBP1 site is thus not restricted to lens cells and follows the general pattern of lens preferred over non-lens cells, and mouse preferred over chicken cells. TheαA-CRYBP1

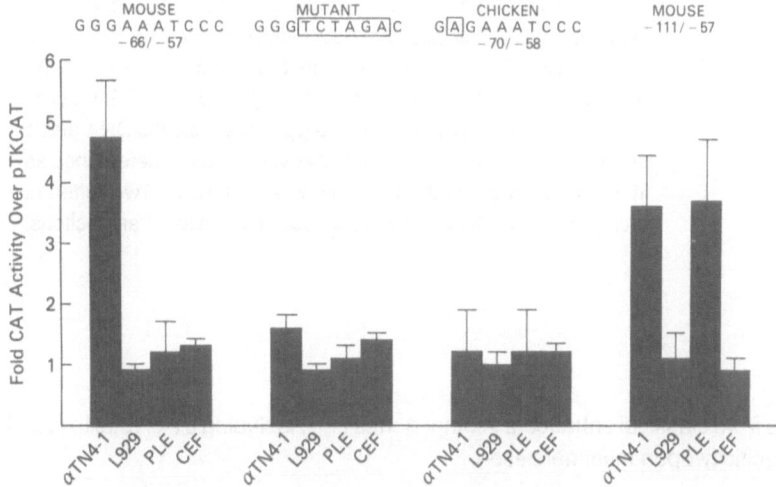

Figure 2. Effects of the αA-CRYBP1 site on transcription. The denoted oligodeoxynucleotides from the mouse or chicken αA-crystallin gene promoters (see Figure 1) were cloned in the forward orientation into pTKCAT (19), and 10 μg of the resulting plasmids co-transfected into αTN4-1, L929, PLE, and CEF cultures with 1-3 μg of pTB1 (5). PLE and CEF primary cultures were established from 14-day embryonic chicken lenses (5, 26) or breast muscle (26), respectively. CEF, αTN4-1, and L929 cells were plated at a density of 5-7x10^5 cells per 60 mm dish one day prior to transfection. PLE cultures consisted of 6 lenses per 60 mm dish. Values shown represent the fold of CAT activity and standard deviation for a test plasmid relative to pTKCAT. CAT activities were standardized for transfection efficiency (fmole of [^{3}H]acetylchloramphenicol produced per minute per unit β-galactosidase). Two different preparations of plasmid DNA were used in two to four transfections.

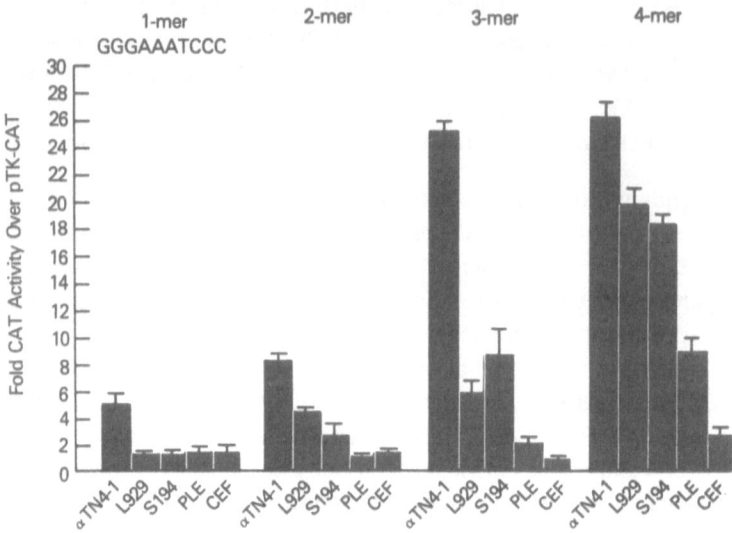

Figure 3. Multimerization of the mouse αA-CRYBP1 site. Oligodeoxynucleotides containing two, three, or four copies of the mouse αA-CRYBP1 (5'-GGGAAATCCC-3') site separated by 5'-AGCT-3', with HindIII compatible ends were synthesized and cloned into HindIII digested pTKCAT at position -130 relative to the tk transcription initiation site. Plasmids were transfected (14) into αTN4-1, L929, PLE, and CEF cultures as described in Figure 2. 1x10⁷ S194 cells (15a) were transfected (25) with 10 μg of test plasmid DNA and 1 μg pTB1 (5). The fold of CAT activity over pTKCAT (determined as in Figure 2) and standard deviations are shown. Two different preparations of plasmid DNA were used in two to four transfections.

site differs from classical enhancers in other genes (11), including δ1-crystallin (13), in its loss of cell-specificity upon multimerization.

Previous northern blot analysis suggested that αA-CRYBP1 is ubiquitous (20), so it was surprising that one copy of the mouse αA-CRYBP1 site did not activate the tk promoter in L929 cells. Preliminary gel mobility shift studies indicate that both αTN4-1 and L929 nuclear extracts produced multiple DNA-protein complexes with the mouse αA-CRYBP1 site (data not shown). The fact that the αA-CRYBP1 site binds L929 nuclear proteins while not activating transcription suggests the existence of lens-specific protein modifications and/or interacting factors. The loss of lens-preference through αA-CRYBP1 site multimerization may arise through a series of protein:protein interactions. Binding of αA-CRYBP1 to a single site in the promoter may not activate transcription in non-lens cells, but binding to multiple sites may compensate for modification and/or additional factor deficiencies and extend expression beyond lens cells. The strong transcriptional activation capabilities of multimerized αA-CRYBP1 sites may provide a mechanism for targeting high level gene expression in a wide variety of eukaryotic cell types and organisms.

ACKNOWLEDGEMENTS

We thank Dr. R. Dubin for technical advice, and D. Chicchirichi for assistance in typing the manuscript. Portions of this manuscript are also presented in Sax, et al., Mol. Cell. Biol. 10: 6813, 1990.

REFERENCES

1. Baldwin, A.S. Jr, K.P. Leclair, H. Singh, and P.A. Sharp. 1990. A large protein containing zinc finger domains binds to related sequence elements in the enhancers of the class I major histocompatibility complex and kappa immunoglobulin genes. Mol. Cell. Biol. 10:1406-1414.

2. Baldwin, A.S., Jr., and P. A. Sharp. 1988. Two transcription factors, NF-kB and H2TF1, interact with a single regulatory sequence in the class I major histocompatibility complex promoter. Proc. Natl. Acad. Sci. USA. 85: 723-727.

3. Baldwin, A.S., Jr., and P. A. Sharp. 1987. Binding of a nuclear factor to a regulatory sequence in the promoter of the mouse H-2K^b class I major histocompatibility gene. Mol. Cell. Biol. 7:305-313.

4. Bohnlein, E., J. W. Lowenthal, M. Siekevitz, D. W. Ballard, B. R. Franza, and W. C. Greene. 1988. The same inducible nuclear proteins regulates mitogen activation of both the interleukin-2 receptor-alpha gene and type 1 HIV. Cell 53:827-836.

5. Borras, T., C.A. Peterson, and J. Piatigorsky. 1988. Evidence for positive and negative regulation in the promoter of the chicken δ1-crystallin gene. Develop. Biol. 127:209-219.

6. Chepelinsky, A.B., C.R. King, P.S. Zelenka, and J. Piatigorsky. 1985. Lens-specific expression of the chloramphenicol acetyltransferase gene promoted by 5' flanking sequences of the murine αA-crystallin gene in explanted chicken lens epithelia. Proc. Natl. Acad. Sci. USA. 82:2334-2338.

7. Chepelinsky, A.B., B. Sommer, and J. Piatigorsky. 1987. Interaction between two different regulatory elements activates the murine aA-crystallin gene promoter in explanted lens epithelia. Mol. Cell. Biol. 7:1807-1814.

8. Clark, L., R. M. Pollock, and R. T. Hay. 1988. Identification and purification of EBP1: a HeLa cell protein that binds to a region overlapping the 'core' of the SV40 enhancer. Genes Develop. 2:991-1002.

9. Earle, W.R., E.L. Schilling, T.H. Stark, N.P. Straus, M.F. Brown, and E. Shelton. 1943. Production of malignancy in vitro. IV. The mouse fibroblast cultures and changes seen in the living cells. J. Natl. Cancer Inst. 4:165-212.

10. Fan, C-M, and T. Maniatis. 1990. A DNA-binding protein containing two widely separated zinc finger motifs that recognize the same DNA sequence. Genes & Develop. 4: 29-42.

11. Gerster, T., P. Matthias, M. Thali, J. Jiricny, and W. Schaffner. 1987. Cell type-specificity elements of the immunoglobulin heavy chain gene enhancer. EMBO J. 6:1323-1330.

12. Gorman, C.M., G.T. Merlino, M.C. Williingham, I. Pastan, and B.H. Howard. 1982. The Rous sarcoma virus long terminal repeat is a strong promoter when introduced into a variety of eukaryotic cells by DNA-mediated transfection. Proc. Natl. Acad. Sci. USA. 79:6777-6781.

13. Goto, K., T.S. Okada, and H. Kondoh. 1990. Functional cooperation of lens-specific and nonspecific elements in the δ1-crystallin enhancer. Mol. Cell Biol.10:958-964.

14. Graham, F.L., and A.J. Van der Eb. 1973. A new technique for the assay of infectivity of human adenovirus 5 DNA. Virology. 52:456-467.

15a. Hyman, R., P. Ralph, and S. Sarkar. 1972. Cell specific antigens and immunoglobulin synthesis of murine myeloma cells and their variants. J. Natl. Cancer Inst. 48:173-184.

15b. Jaworski, C.J., A.B. Chepelinsky, and J. Piatigorsky 1991. The αA-crystallin gene: Conserved features of the 5'-flanking regions in human, mouse, and chicken, J. Mol. Evol. 33:495-505.

16. Klement, J.F., E.F. Wawrousek, and J. Piatigorsky. 1989. Tissue-specific expression of the chicken αA-crystallin gene in cultured lens epithelia and transgenic mice. J. Biol. Chem. 264:19837-19844.

17. Lenardo, M.J., and D. Baltimore. 1989. NF-kB: A pleiotropic mediator of inducible and tissue-specific gene control. Cell. 58:227-229.

18a. Matsuo,I.,M. Kitamura, K. Okazaki, and K. Yasuda. 1991. Binding of a factor to an enhancer element responsible for the tissue-specific expression of the chicken αA-crystallin gene. Develop. 113:539-550.

18b. Maekawa, T.,H. Sakura, T. Sudo, and S. Ishii. 1989. Putative metal finger structure of the human immunodeficiency virus type 1 enhancer binding protein HIV-EPI. J. Biol. Chem. 246:14591-14593.

19. Miksicek, R., A. Heber, W. Schmid, U. Danesch, G. Possechert, M. Beato, and G Schutz. 1986. Glucocorticoid responsiveness of the transcriptional enhancer of moloney murine sarcoma virus. Cell 46: 283-290.

20. Nakamura, T., D.M. Donovan, K. Hamada, C.M. Sax, B. Norman, J.R. Flanagan, K. Ozato, H. Westphal, and J. Piatigorsky. 1990. Regulation of the mouse αA-crystallin gene: Isolation of a cDNA encoding a protein that binds to a cis sequence motif shared with MHC class I and other genes. Mol. Cell. Biol. 10:3700-3708.

21. Neumann, J.R., C.A. Morency, and K.O. Russian. 1987. A novel rapid assay for chloramphenicol acetyltransferase gene expression. BioTech. 5:444-447.

22. Nielsen, D.A., J. Chou, A. J. Mackrell, M.J. Casadaban, and D.F. Steiner. 1983. Expression of a preproinsulin-β-galactosidase gene fusion in mammalian cells. Proc. Natl. Acad. Sci. USA. 80:5198-5202.

23. Overbeek, P.A., A.B. Chepelinsky, J.S. Khillan, and J. Piatigorsky. 1985. Lens-specific expression and developmental regulation of the bacterial chloramphenicol acetyltransferase gene driven by the murine αA-crystallin promoter in transgenic mice. Proc. Natl. Acad. USA. 82:7815-7819.

24. Piatigorsky, J. 1989. Lens crystallins and their genes: diversity and tissue-specific expression. FASEB J. 3:1933-1940.

25. Sambrook, J., E.F. Fritsch, T. Maniatis. 1989. Transfection using DEAE-dextran: Protocol I. p. 16.42-16.44. In Molecular Cloning. A Laboratory Manual. Cold Spring Harbor Laboratory Press, Cold Spring Harbor, NY.

26. Sax, C.M., F.X. Farrell, Z.E. Zehner, and J. Piatigorsky. 1990. Regulation of vimentin gene expression in the ocular lens. Develop. Biol. 139:56-64.

27. Sommer, B., A.B. Chepelinsky, and J. Piatigorsky. 1988. Binding of nuclear proteins to promote elements of the mouse αA-crystallin gene. J. Biol. Chem. 263: 15666-15672.

28. Wawrousek, E.F., A.B. Chepelinsky, J.B. McDermott, and J. Piatigorsky. 1990. Regulation of the murine αA-crystallin promoter in transgenic mice. Develop. Biol. 137: 68-76.

29. Wirth, T. and D. Baltimore. 1988. Nuclear factor NF-KB can interact functionally with its cognate binding site to provide lymphoid-specific promoter function. EMBO J. 7: 3109-3113.

30. Wistow, G.J., and J. Piatigorsky. 1988. Lens crystallins: The evolution and expression of proteins for a highly specialized tissue. Ann. Rev. Biochem. 57:479-504.

31. Yamada, T., T. Nakamura, H. Westphal, and P. Russell. 1990. Synthesis of α-crystallin by a cell line derived from the lens of a transgenic animal. Curr. Eye Res. 9:31-37.

32. Yano, O., J. Kanellopoulos, M. Kieran, O. Le Bail, A. Isreal, and P. Kourilsky. 1987. Purification of KBF1, a common factor binding to both H-2 and β2-microglobulin enhancers. EMBO J. 6:3317-3324.

RECENT ADVANCE ON THYMIDYLATE SYNTHASE-DIHYDROFOLATE REDUCTASE FROM *Plasmodium falciparum*

Worachart Sirawaraporn

Department of Biochemistry, Faculty of Science
Mahidol University, Bangkok 10400, Thailand

INTRODUCTION

The bifunctional TS-DHFR of *Plasmodium falciparum* is one of the few drug targets for malaria. The antifolate drugs such as pyrimethamine and proguanil are potent inhibitors of *Plasmodium* DHFR and have been widely used for malaria chemotherapy for more than two decades (1,2). However, the emergence of antifolate resistant parasites in many areas of the world led to the interest in the mechanisms of drug resistance which would require knowledge of the structure of the target enzyme. Unlike antifolate drug resistance in *Leishmania* (3,4), the mechanism of resistance in *Plasmodium* has been suggested to be due to a structurally altered enzyme (5-10). Recently, the gene encoding the bifunctional enzyme from *P. falciparum* has been cloned and isolated from genomic library (11-14). Evidence from the analysis of the DHFR portion of gene obtained from a number of pyrimethamine-resistant strains further supported point mutations of the gene as a mechanism of pyrimethamine resistance (12-15). Thus far, point mutation at amino acid residue 108 of DHFR domain was proposed to be associated with pyrimethamine resistance in *P. falciparum*. Additional mutations at some other amino acid residues were also reported to involve the increase in level of resistance (13,15). However, the study of the role of these amino acids in pyrimethamine resistance has been impeded by the limited quantities of enzyme obtained from *in vitro* cultured parasite, though the purification scheme has been reported (16). We describe here the construction of TS-DHFR recombinant plasmid and heterologous expression of the first recombinant malarial enzyme in *E. coli*. We also describe the method by which the recombinant plasmid was genetically manipulated to express different levels of pyrimethamine resistant enzymes. Analysis of the protein products of the genes with regards to their kinetic properties and inhibition by pyrimethamine provides direct evidence of point mutations of the gene as a mechanism of pyrimethamine resistance in *P. falciparum*.

MATERIALS AND METHODS

Materials

pGem3 plasmid containing 4.9 kb Xba1 fragment of TS-DHFR (pGem3 Xba1) and

Biotechnology and Environmental Science: Molecular Approaches
Edited by S. Mongkolsuk et al., Plenum Press, New York, 1992

pUC13 plasmid containing 2 kb EcoR1 fragment (pUC13 5'R1) for moderately pyrimethamine-resistant, strain HB3, and amplified gene fragments of wild type strain 3D7 and pyrimethamine-resistant, strain 7G8 were isolated as previously described (12). Restriction enzymes, T_4-DNA polymerase, and T_4-DNA ligase were purchased from New England Biolabs. Deoxyadenosine 5'-(α-[^{35}S] thio) triphosphate (>1,000 Ci/mmol) and [^{3}H] FdUMP (20 Ci/mmol) were from Amersham Corp. and Moravek Biochemicals, respectively. M13 mp18 and mp19 RF DNA were products from Pharmacia. The *E. coli* K12 strain χ2913 (thy A572) was a gift from Dr. R. Thompson (University of Glasgow). Pure agarose and low melting temperature agarose were obtained from Bethesda Research Laboratories. Oligonucleotides were synthesized by phosphoramidite method (17-19) using an automated DNA synthesizer (Applied Biosystem) at UCSF Biomolecular Resource Center and purified by NENSORB cartridge from NEN. The sequenase DNA sequencing kit was purchased from USB. 10 formylfolate was prepared from folic acid as described (20) and coupled to aminohexyl-Sepharose CL-6B prepared according to method as described (21). FAH_2 was prepared from folic acid (22). All other reagents were the purest commercially available.

DNA Manipulations

The general methods for ligation (23), transformation of *E.coli* (24), and rapid alkali preparation of plasmid DNA (25) were performed as described. Purification of DNA fragments was achieved either by using low melting temperature agarose (26) or electroelution. Ligation of adapter to purified restriction fragments and vectors was performed as described (27). For fragment ligation of 6 synthetic oligonucleotides coding for the first 47 amino acid of TS-DHFR gene into a modified pUC18 plasmid, approximately 2.5 pmoles of the complementary strand of synthetic oligonucleotide was annealed and followed by ligation with the endonuclease-digested vectors at molar ratios of 5:1 and 10:1. The DNA sequences of the gene constructs were initially confirmed by restriction analysis and subsequently verified by dideoxy method of DNA sequencing (28).

Growth Conditions

Bacterial clones were grown in LB containing 50 mg/mL ampicillin at 37°C with shaking. To obtain maximum expression of both TS and DHFR, the bacteria culture of 1 % inoculum size was harvested at 12 hr after inoculation. Cell extract was then analysed for DHFR and TS activity.

Preparation of cell extracts

Cell paste from 4-5 L of *E.coli* χ2913 cells (thy⁻) haboring pTDSD (3D7), pTDSD (HB3) and pTDSD (7G8) were thawed and resuspended in 40 mL of buffer A (50 mM Tris, 1 mM EDTA , 5 mM 2-mercaptoethanol, pH 7.5) with a cocktail of six protease inhibitors (29). The cell suspension was sonicated on ice using Branson sonifier Model 350 equipped with standard 3/4" horn. Sonication was performed for 3 min at 50 % duty pulse setting with 60-70 watt output. The completeness of cell disruption was monitored by microscopic examination. Clear extracts were obtained after centrifugation at 30,000 g for 1 hr at 4°C.

Protein analysis

Protein was determined by the method of Read and Northcote (30). Electrophoresis of proteins on 12% SDS-PAGE followed by staining with Coomassie Blue R250 was performed as described (31). Western blotting of proteins onto Immobilon-P membrane (Millipore) and sequencing of interested protein from excised membrane (32) were performed using an ABI 470 A protein sequencer/120 A PTH analyzer at UCSF Biomolecular Resource Center. The activities of DHFR and TS were determined spectrophotometrically as described elsewhere for the *Leishmania* enzyme (29). One unit of enzyme activity is defined as the amount of enzyme required to reduce 1 nmole of 7,8 FAH_2 to FAH_4 per min at 25°C. Analysis of *Plasmodium* TS-DHFR subunit was performed by the formation of the covalent complex of [^{3}H]FdUMP, CH_2FAH_4 and enzyme (33) followed by SDS-PAGE on 12% polyacrylamide gels. The Coomassie-stained gel was soaked with autoradiography enhancer (Enlightning™, DuPont) for 15 min prior to drying the gel. The gel was exposed on KODAK X-OMAT film at -80°C for 48 h prior to film development.

Analysis of catalytically active *Plasmodium* TS-DHFR with regard to its native molecular weight was performed by gel filtration using FPLC equipped with Superose 12™ column (Pharmacia). The sonicated-crude extract (1 mL, ~12 mg/mL) in 50 mM potassium phosphate, pH 7.2 containing 0.15 M NaCl was injected into the column pre-equilibrated with the same buffer at a flow rate of 0.2 mL/min. The column was separately calibrated with the following protein standards: Catalase (232 kDa), bovine serum albumin (66.2 kDa), ovalbumin (43 kDa), chymotrypsinogen A (25 kDa), and ribonuclease A (13.7 kDa).

Protein purification

Affinity chromatography

Crude extract (~50 mL) was passed through 10 formylfolate-Sepharose column (1.0 X 4.0 cm) pre-equilibrated with buffer A at a flow rate of 0.5 mL/min. The column was washed with the same buffer containing 0.45 M NaCl (buffer B) at a flow rate of 1 mL/min until protein was undetectable in the effluent (~80 mL). 7,8 FAH_2 (4 mM) in buffer B (20 mL) was then applied onto the column at a flow rate of 0.5 mL/min. The enzyme was eluted from the column in the first 15 mL. Active fractions (5 mL) were collected, pooled, dialyzed/concentrated against buffer A at 4°C using a negative pressure micro protein dialysis/concentrator (Micro-ProDiCon, Bio-Molecular Dynamics).

Anionic exchange chromatography

The concentrated sample (1 mL, ~ 0.1 mg. protein) from affinity chromatography was injected onto Mono Q HR5/5 pre-equilibrated with buffer A. The column was washed with buffer A at a flow rate of 0.2 mL/min for 5 mL before starting with two-step gradient of NaCl using buffer A containing 1 M NaCl, i.e. 0 to 0.2 M NaCl (30 mL), 0.2 M to 1 M NaCl (10 mL). The TS-DHFR was eluted from the column at 70-90 mM NaCl. Active fractions (5 mL) were pooled and concentrated against buffer A at 4°C using a negative pressure micro protein dialysis/concentrator to a final volume of 0.5-1 mL.

Complementation of TS-Deficient E. coli

Plasmids pTDSD (3D7), pTDSD (HB3), pTDSD (7G8) and pUC18 (as a control) were transformed into the TS-deficient *E.coli* K12 strain χ2913 as described (24). The transformation mixtures were plated on minimal agar plates (34) containing 50 ug/mL ampicillin, with and without 50 ug/mL thymidine. Colonies which express active TS can be visualized from the plates lacking thymidine after incubation at 37°C overnight.

RESULTS

Cloning the coding sequence

The 1,824 bp coding sequence of *P. falciparum* TS-DHFR was assembled with a synthetic oligonucleotide at the 5' end (nt 1-138), a 243 bp genomic Nco1-Xba1 fragment from pUC13 (5'R1) (nt 139-382), a 1427 bp genomic Xba1-Bbv1 fragment from pGem3 (Xba1) (nt 383-1810), and a synthetic oligonucleotide at the 3' end (nt 1811-1824).

Construction of expression vectors

A 16/24 bp synthetic oligonucleotide containing a ribosomal binding sequence (ON 4, Table1) was introduced at the Sst1/BamH1 sites of parent plasmid pTDPF to give pTDSD,

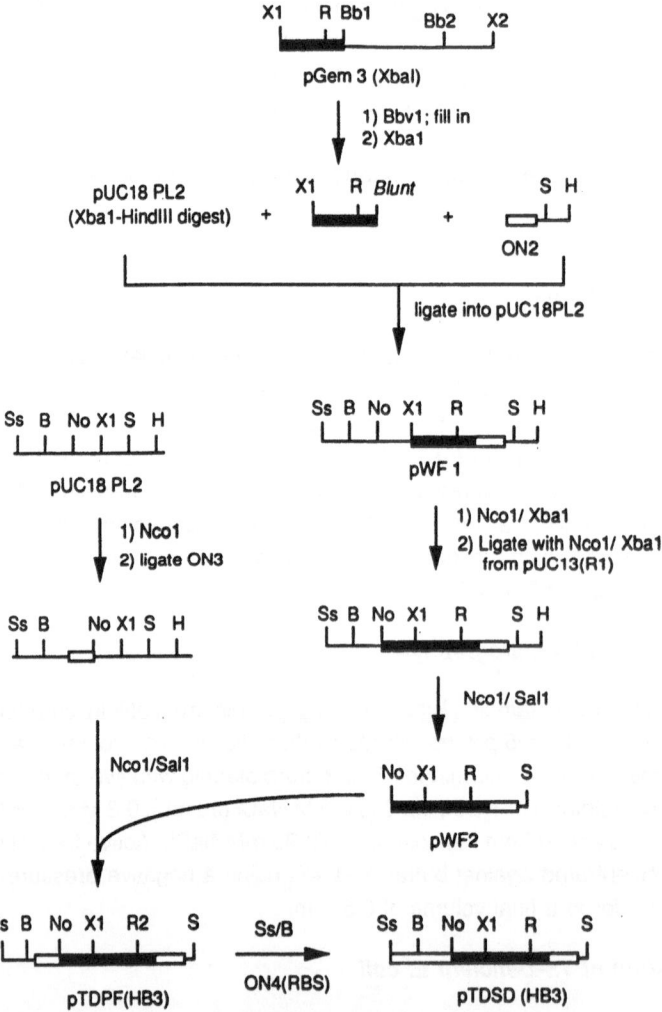

Figure 1 Construction of the expression vector pTDSD (HB3) containing *P. falciparum* TS-DHFR gene. The solid bars represent sequences derived from genomic DNA, and open bars are sequences from the synthetic oligonucleotides (ON 2 and ON 3). Restriction endonuclease sites are Ss for Sst1, K for Kpn1, B for BamH1, No for Nco1, X for Xba1, S for Sal1, H for HindIII, R for EcoR1, Bb for Bbv1.

Table 1 Synthetic Oligonucleotide Duplexes Used in Constuction of the *P. faclciparum* TS-DHFR[a] Expression Vectors

Oligonucleotide	Sequence
Polylinker duplex (ON1)	5'-CGGATCCCATGGTCTCTAGATCGTCGACA 3'
	3'-CATGGCCTAGGTACCAGATCTAGCAGCTGTTCGA 5'
3' end (ON2)	5'-TGGATATGGCTGCTTAAGATATCGTCGACA 3'
	3'-ACCTATACCGACGAATTCTATAGCAGCTGTTCGA 5'
5' end (ON3)[a]	
ON3A	5'-CATGATGGAACAAGTCTGCGACGTTTCGATATCTATGCCATATGTGCAT 3'
ON3B	3'-TACCTTGTTCAGACGCTGCAAAAGCTATAGATACCGGTATACACGTACAACATTCC 5'
ON3C	5'-GTTGTAAGGTTGAAAGCAAAAATGAGGGGAAAAAATGAGGTTTTTAAT 3'
ON3D	3'-AACTTTCGTTTTTACTCCCCTTTTTTTTACTCCAAAAATTATTGATGT 5'
ON3E	5'-AACTACACATTTAGAGGTCTAGGAAATAAAGGAGTATTAC 3'
ON3F	3'-GTAAATCTCCAGATCCTTTATTTCCTCATAATGGTAC 5'
Ribosome binding site (ON4)	5'-CGTAAGAGGAGTTAAA 3'
	3'-TCGAGCATTCTCCTCAATTTCTAG 5'

[a] ON3A-F were ligated to form ON3, as described in the text.

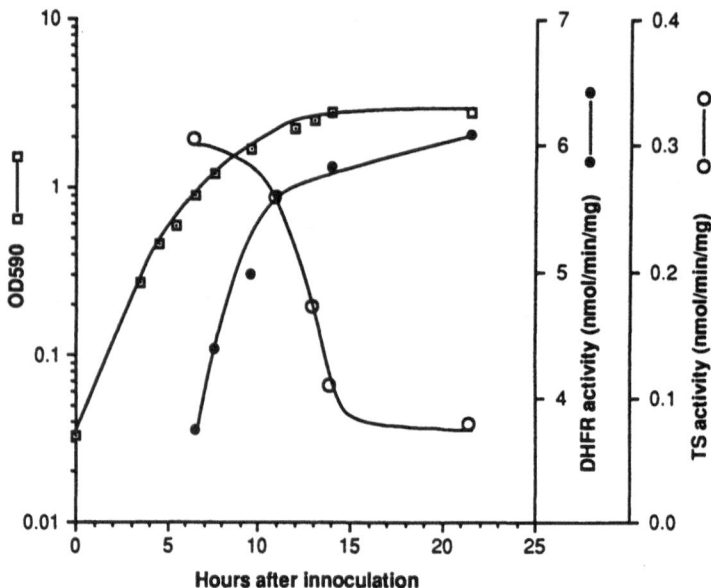

Figure 2 Expression of *Plasmodium* TS-DHFR in *E. coli* χ2913 harboring pTDSD (HB3). 10 mL of an overnight culture of *E. coli* χ2913 harboring pTDSD (HB3) was inoculated into 1 L minimum media supplemented with 50 ug/mL Ampicillin. Growth of the bacteria was monitored by absorbance at 590 nm over a period of 22 hr. Aliquots of the culture (50 mL) was collected intervally and centrifuged to pellet the cells. The activity of DHFR and TS was determined as described in Materials and Methods. , □——□ OD 590; ●——● , DHFR activity (nmol/min/mg); O——O , TS activity (nmol/min/mg).

such that the RBS is separated from the start codon by 10 bases. Figure 1 illustrates the construction of parent plasmid pTDPF and expression plasmid pTDSD. Since the Nco1 - Xba1 fragment (243 bp) of the DHFR gene includes point mutations thus far linked to pyrimethamine resistance, recombinant plasmids of different levels of pyrimethamine resistance could simply be constructed by cassette mutagenesis using Nco1-Xba1 fragments from the corresponding cloned TS-DHFR genes, i.e., the Nco1-Xba1 fragment from 3D7 was introduced into the corresponding sites of pTDSD(HB3) to give pTDSD(3D7) which contains the wild type Ser 108 instead of the Asn in HB3. Likewise, the Nco1-Xba1 fragment from 7G8 was cloned into pTDSD(HB3) to give pTDSD(7G8) which contains Ile 51 and Asn 108 instead of the wild type Asn 51 and Ser 108.

Expression of recombinant TS-DHFR

In initial studies of expression of TS-DHFR, we examined several different *E. coli* strains as hosts for pTDSD(HB3), pTDSD(7G8) and pTDSD(3D7). (Table 2). All three of the constructs complemented growth of TS-deficient *E. coli* χ2913, indicating that catalytically active TS was being expressed from the plasmids. After correcting for host enzyme, the DHFR activities in transformed cells were 3 to 10-fold higher than found in the host cell. The TS activities were erratic, as indicated by the ratios of DHFR/TS. We attribute this to proteolysis or other forms of degradation of TS which diminish its activity.

The TS-DHFR activities and growth of *E. coli* χ2913 harboring pTDSD (HB3) in minimum media with 50 ug/mL ampicillin was observed over a period of 22 hr (Fig. 2). The activity of DHFR followed cell growth and leveled after about 12 hr, whereas TS activity sharply declined after about 7 hr. At 6.5 hr the DHFR/TS ratio was ~12, and the ratio increased dramatically with time due to the diminished TS activity. By 22 hr after inoculation, the DHFR

Table 2 TS and DHFR Activities in Crude Extracts of pTDSD-Transformed E. *coli* Hosts[a]

host	plasmid	condition	DHFR (nmol min^{-1} mg^{-1})	TS (nmol min^{-1} mg^{-1})
TB1		LB	2.4	0.13
TB1	pTDSD(HB3)	LB+Amp	10.1	0.55
HB101		LB	1.2	0.50
HB101	pTDSD(HB3)	LB+Amp	10.5	0.60
χ2913		LB	2.5	ND
χ2913	pTDSD(HB3)	LB+Amp	16.2	0.18
χ2913		minimum + Thd	3.7	ND
χ2913	pTDSD(HB3)	minimum + Amp	13.0	0.15
χ2913	pTDSD(HB3)	minimum + Thd + Amp	18.2	0.18
χ2913	pTDSD(3D7)	minimum + Amp	6.6	0.21
χ2913	pTDSD(7G8)	minimum + Amp	6.0	0.05
D3-157		LB	ND[b]	0.37
D3-157	pTDSD(HB3)	LB + Amp	20.4	0.86

[a] All cultures were grown for about 12 h to stationary phase. [b] Not detectable.

activity was approaching its maximum level but the TS activity was only ~25% of the activity at 6.5 hr. with a DHFR/TS ratio of ~77. Loss of TS activity was previously observed with *Leishmania* TS-DHFR, and shown to be a result of proteolysis (35). With this in mind, we transformed the protease deficient cell lines *E.coli* CAG 456 and FB 974 (36) with pTDSD (HB3), but enzyme activity was not detected in crude extracts of these strains (data not shown).

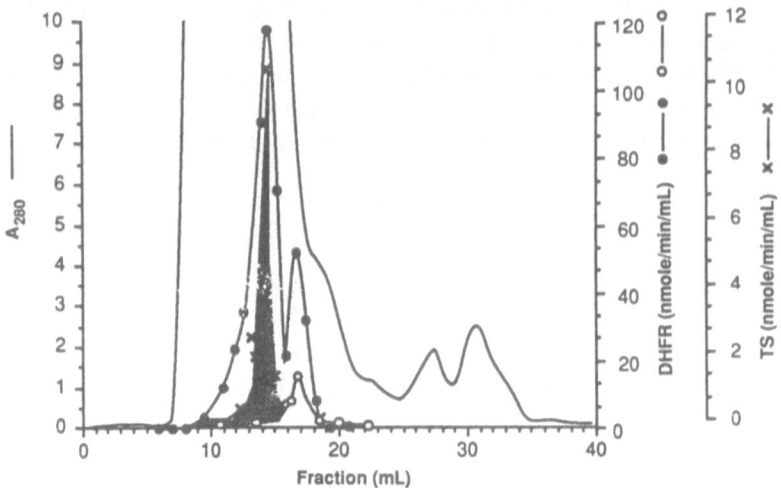

Figure 3 Superose   12 HR chromatography profiles of crude extract of *E. coli* χ2913 harboring plasmid pTDSD (HB3). ——— , A280; ○———○ , DHFR activity from *E. coli* χ2913 (light shading); ●———● , DHFR activity from *E. coli* χ2913 harboring plasmid pTDSD (HB3); x———x , TS activity (dark shading) co-elutes with DHFR peak of pTDSD (HB3). The makers used (Mr; Ve/Vo) were catalase (232 kDa; 1.30), BSA (67 kDa; 1.47), ovalbumin (43 kDa; 1.54), chymotrypsinogen (25 kDa; 1.69), and ribonuclease (13.7 kDa; 1.77).

Superose 12 gel filtration of extracts of   χ2913 harboring pTDSD (HB3) showed that the TS and DHFR activities co-eluted as a single peak of about 110 ± 15 kDa, as compared with the calculated size of 143 kDa (Fig. 3);   the recovery was about 40% and there were 116 U DHFR and 3.1 U TS. The small trailing peak of DHFR activity with Mr ~20 kDa was probably host *E. coli* DHFR ; indeed a control extract from the host χ2913 showed a small DHFR peak eluting with Mr 20,000. Moreover, when crude extract of χ2913 harboring pTDSD (HB3) was treated with CH_2FAH_4 and $[6-^3H]$ FdUMP to form the enzyme-CH_2FAH_4-$[6-^3H]$ FdUMP covalent complex and analyzed by SDS-PAGE, there was a band with subunit Mr of 67 kDa (Fig. 4, lane 4) upon autoradiography with no indication of host TS (Mr= 20,000). However, there were several bands with Mr ~35-40 kDa (Fig. 4, lane 5) indicating proteolysis. Affinity chromatography on 10-formylfolate Sepharose clearly showed homogeneous product with Mr~67 kDa (Fig.4, lane 6). When treated similarly, the control extract from the host χ2913 showed no radioactive enzyme complex upon autoradiography (data not shown).

Purification and characterization

The expressed TS-DHFR from *E.coli* χ2913 harboring pTDSD (HB3) was purified to homogeneity using a combination of affinity chromatography on 10-formylfolate Sepharose and anion exchange chromatography (Fig.5A and 5B). After affinity chromatography the preparation showed one major band on SDS-PAGE migrating with mass of 67 kDa and several minor bands of 40-50 kDa and 25 kDa (Fig. 4, land 2). Ternary complex of the enzyme with CH_2FAH_4 and $[^3H]$ FdUMP clearly showed that the 67 kDa and 40 kDa proteins had TS active site. Further purification of the enzyme by Mono Q resulted in homogeneous TS-DHFR with molecular mass of 67 kDa as shown by single band on SDS-PAGE (Fig. 4, lane 3) and single band on autoradiogram of the ternary complex of the enzyme with CH2FAH4 and [3H] FdUMP (Fig. 4, lane 6). The 67 kDa protein was eluted at 70-90 mM NaCl, corresponding to fractions 16-21 in Fig. 5B, whereas host DHFR was eluted from Mono Q at 600-700 mM NaCl (data not shown).

In order to further verify that the 67 kDa band on SDS-PAGE was attributed to *Plasmodium* TS-DHFR, the proteins were electrotransferred onto Immobilon-P membrane and the excised band was subjected to direct micro sequencing analysis. Ten cycles of analysis gave the sequence M M E Q V C D V F D which perfectly matched the nucleotide sequences reported for *Plasmodium* TS-DHFR gene.

Table 3 Kinetic Parameters of Purified Recombination *P. falciparum* TS-DHFR[a]

parameter	3D7 Asn 51 Ser 108	HB3 Asn 51 Asn 108	7G8 Ile 51, Asn 108
K_m for H_2folate (μM)	1.0 ± 0.1	11.0 ± 1	5.8 ± 0.5
K_m for NADPH (μM)	4.2 ± 0.2	1.1 ± 0.2	9.7 ± 3.6
K_i for pyrimidine (nM)	0.10 ± 0.03	2.0 ± 0.4	230 ± 50

[a] Conditions used are described under Materials and Methods.

Kinetic parameters

The kinetic parameters of the purified recombinant TS-DHFR enzymes are given in Table 3. The K_m values for substrates FAH_2 and NADPH are within the range reported previously from the enzymes obtained from natural sources (37). The K_i values for Pyr of the wild type (3D7) and moderately resistant (HB3) enzyme were also comparable to those reported previously (37). For 7G8 recombinant enzyme, however, the K_i for Pyr is ~2,300 fold higher than that of wild type enzyme, a value which is ~26 fold higher than that reported from natural sources (37). The above data suggest that amino acid changes at position 51 and 108 are responsible for conferring pyrimethamine resistance in malaria parasites.

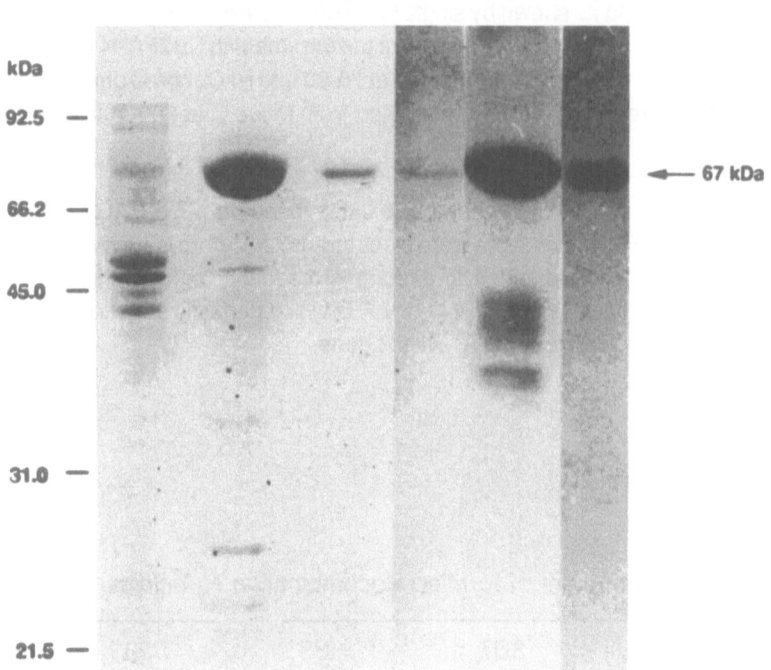

Figure 4 SDS-PAGE of TS-DHFR treated with [³H] FdUMP and CH_2FAH_4. Lane 1-3 were stained with Coomassie blue-R250; lanes 4-6 are autoradiograms of lanes 1-3, respectively. Lanes 1 and 4, crude extract from *E. coli* χ2913 harboring plasmid pTDSD (HB3); lanes 2 and 5, after 10 formylfolate Sepharose column; lanes 3 and 6, after mono Q FPLC. Molecular weight markers were phosphorylase b (97.4 kDa), bovine serum albumin (66.2 kDa), ovalbumin (45.0 kDa), carbonic anhydrase (31.0 kDa), soybean trypsin inhibitor (21.5 kDa) and lysozyme (14.4 kDa). The arrow shows the 67 kDa Plasmodium TS-DHFR.

DISCUSSION

The coding sequence of moderately resistant *P. falciparum* was assembled in an expression vector using a combination of synthetic oligonucleotides and gDNA. The vector contains a RBS 10 bases from the initiation codon, and several unique restriction sites. In particular, unique Nco1 and Xba1 sites separate a 243 bp Nco1-Xba1 fragment (nt 139-382) which codes for all of the mutations thus far reported to be associated with pyrimethamine resistance. Thus, expression vectors containing any of these mutations can readily be constructed by cassette mutagenesis of the appropriate Nco1-Xba1 fragment from *P. falciparum* isolates. Using this approach, we have vectors expressing TS-DHFR from wild type strain 3D7, moderately pyrimethamine-resistant strain HB3 which has a Ser to Asn mutation at amino acid residue 108, and highly resistant strain 7G8 which has double mutation of Asn to Ile at residue 51 and Ser to Asn at residue 108.

The successful expression of *P. falciparum* TS-DHFR in *E. coli* harboring these vectors was unequivocally demonstrated. First, transformation of a TS-deficient *E. coli* host resulted

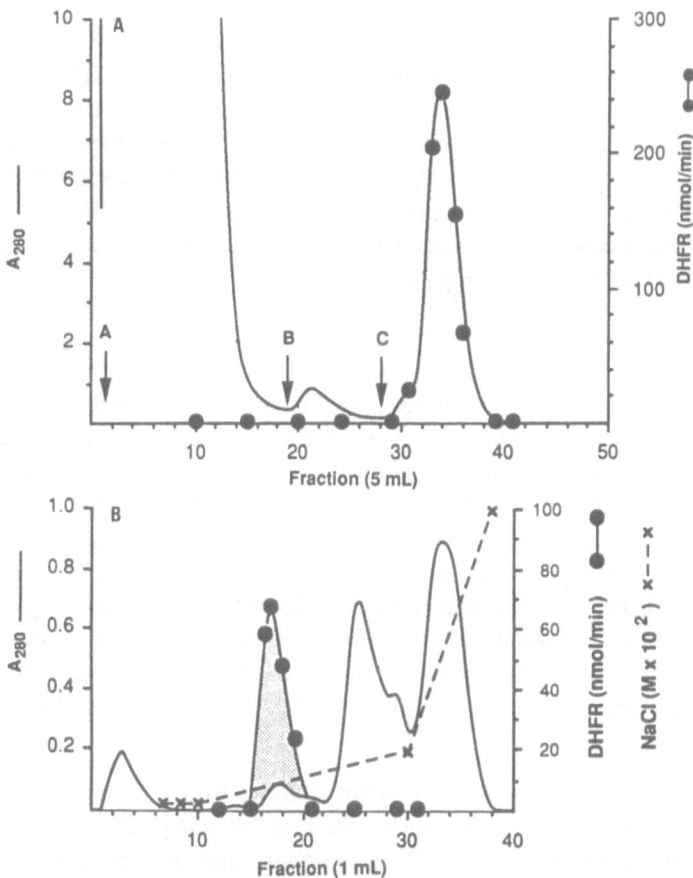

Figure 5 Purification of recombinant TS-DHFR on (A) 10 formylfolate-Sepharose CL-6B (B) anionic exchange Mono Q. Crude extract was prepared as described in Materials and Methods and circulated on 10 formylfolate-Sepharose column. Buffer A is 50 mM Tris, 1mM EDTA, 5 mM 2-mercaptoethanol, pH 7.5; buffer B is buffer A containing 0.45 M NaCl; buffer C is buffer B containing 4 mM FAH_2.. ——— ., A280; ●———● , DHFR activity (nmole/min/ml); x———x , NaCl concentration (M).

in genetic complementation of the deficiency. Second, gel filtration showed that both enzyme activities comigrated with Mr of 110,000, in agreement with the value reported for the enzyme from natural sources (16,38). The expected Mr of the native bifunctional protein is 120,000, whereas host DHFR and TS have Mr values of 20,000 and 70,000, respectively. Third, the denatured covalent TS-FdUMP-CH_2FAH_4 complex migrates on SDS-PAGE with a Mr of 67,000, in accord with the subunit size of the bifunctional enzyme. Finally, N-terminus sequencing of the first 10 amino acids from the purified 67 kDa subunit perfectly matched the reported *P. falciparum* TS-DHFR nucleotide sequence.

The expressed TS-DHFR was not stable, particularly the activity of TS decreased dramatically after 22 hr culture (Fig.2). To obtain maximum activity of both enzymes, we harvested the culture at 12 hr after inoculation (culture of 1% inoculum size). This is equivalent to OD_{590} ~2.2-2.5 or late log phase. Since *L. major* TS-DHFR is sensitive to proteases (35) and degradation of TS-DHFR in *P. falciparum* extracts has been reported (16), we therefore expected the proteolysis to occur in our system. As a precautionary measure, all extracts were prepared in the presence of a mixture of protease inhibitors (29), and processed immediately to minimize proteolysis. Despite these precautions, we observed degraded products of 38-45 kDa (Fig.4).

An attempt to purify the recombinant *Plasmodium* TS-DHFR using MTX-Sepharose as in the case of recombinant *Leishmania* TS-DHFR failed. The enzyme bound very tightly to the affinity matrix and could not be eluted from the column even with high concentration of substrate, a result confirming earlier reports (16, 38). In initial experiments, hydroxyapatite (HTP) was used as the first purification step to remove host DHFR from *Plasmodium* TS-DHFR since the former did not bind HTP and was eluted in the washthrough step (data not shown). In subsequent experiments, this step was omitted to simplify the protocol of purification. An affinity column on 10 formylfolate-Sepharose followed by anionic exchange column on Mono Q FPLC resulted in pure 67 kDa band on SDS-PAGE and autoradiogram (Fig.4, land 3 and 6). Degradation of enzyme seems to be a major problem during purification as shown by the increased intensity from autoradiogram of degradative complex of Mr ~40 kDa after affinity chromatography (Fig. 4, lane 5). The degradative product was removed by anionic exchange Mono Q FPLC (Fig. 4, lane 6).

The purified recombinant enzymes showed kinetic properties similar to the corresponding authentic enzymes except for the K_i for Pyr from pTDSD (7G8) which is ~26 fold higher than the reported value (37). It is unlikely that contamination of host DHFR contributes to an apparent high K_i since the proteins were eluted at different NaCl concentrations. With the available expression plasmids, it is therefore feasible to construct plasmids which express enzyme of different levels of Pyr resistance. Moreover, the availability of recombinant enzymes may also provide the opportunities to to study the enzyme structure, function and design to more selective inhibitors.

SUMMARY

The coding sequence of bifunctional thymidylate synthase-dihydrofolate reductase (TS-DHFR) from moderate pyrimethamine resistant strain of *Plasmodium falciparum* was assembled in a pUC expression vector. Insertion of a synthetic oligonucleotide duplex containing a ribosomal binding sequence at 5' upstream to ATG start codon resulted in a recombinant plasmid which expressed catalytically active TS-DHFR in *E. coli* . The expression of Plasmodium enzyme was demonstrated as follows: First, transformation of a TS-deficient *E. coli* resulted in genetic complementation of the deficiency and the activities of TS and DHFR were significantly higher than those in control host extracts. Second, TS and DHFR activities copurified upon chromatography with an apparent molecular weight ~110 kDa, a characteristic of protozoan bifunctional enzyme. Third, the enzyme formed a covalent complex with [^{3}H] FdUMP-CH_2H_4folate, and SDS polyacrylamide gel electrophoresis showed the subunit of the recombinant enzyme to be 67 kDa. Finally, N-terminus sequencing of the 67 kDa protein indicated that the N-terminus was not blocked and the sequences of the first 10 amino acids perfectly matched the reported nucleotide sequence. The recombinant enzyme was purified to homogeneity using 10 formylfolate affinity chromatography followed by FPLC on Mono Q column. Since the coding sequence possesses unique Nco1 and Xba1 sites which flank 243 bp of the DHFR gene that include point mutations thus far linked to pyrimethamine resistance, recombinant plasmids of wild type (3D7) and highly pyrimethamine resistant (7G8) gene were constructed by cassette mutagenesis using Nco1-Xba1 fragments from the corresponding cloned TS-DHFR genes. Characterization of the purified-recombinant enzymes with regard to their kinetic properties and inhibition by pyrimethamine strongly clearly indicated that point mutations are the molecular basis of pyrimethamine resistance in *P. falciparum*.

ABBREVIATION

TS-DHFR, thymidylate synthase-dihydrofolate reductase; FAH_2, 7,8-dihydrofolate; FAH_4, tetrahydrofolate; CH_2FAH_4, 5,10-methylenetetrahydrofolate; [6-^{3}H]FdUMP, 5-fluoro-2'-deoxy[6-^{3}H]uridine 5'-monophosphate; SDS-PAGE, sodium dodecyl sulfate-polyacrylamide gel electrophoresis; dNTP, deoxynucleotide triphosphate; EDTA, ethylenediaminetetraacetic acid; ON, oligonucleotides; nt, nucleotide; RBS, ribosomal binding sequence

ACKNOWLEDGEMENTS

This work was supported by TDR/RF joint venture programme.

REFERENCES

1. Hitchings, G. H. (1952) *Trans. R. Soc. Trop. Med. Hyg.* **46**, 467-473.
2. Ferone, R., Burchall, J.J. & Hitching, G.H. (1969) *Mol. Pharmacol.* **5**, 49-59.
3. Coderre, J.A., Beverley, S.M., Schimke, R.T. & Santi, D.V. (1983) *Proc. Natl. Acad. Sci. USA* **80**, 2132-2136.
4. Beverley, S.M., Corderre, J.A., Santi, D.V. & Schimke, R.T. (1984) *Cell* **38**, 431-439.
5. Ferone, R. (1970) *J. Biochem.* **245**, 850-854.
6. Sirawaraporn, W., & Yuthavong, Y. (1984) *Mol. Biochem. Parasitol.* **10**, 355-367.
7. McCutchan, T.F., Welsh, J.A., Dame, J.B., Quakyi, I.A.,.Graves, P.M., Drake, J.C. & Allegra, C.J. (1984) *Antimicrob.Agents Chemother.***26**, 656-659.
8. Walter, R.D. (1986) *Mol. Biochem. Parasitol.* **19**, 61-66.
9. Banyal, H.S. & Inselburg, J. (1986) *Exp. Parasitol.* **62**, 61-70.
10. Inselburg, J., Bzik, D.J. & Horii, T. (1987) *Mol. Biochem. Parasitol.* **26**, 121-134.
11. Bzik, D.J., Li, W., Horii, T. & Inselburg, J. (1987) *Proc. Natl. Acad. Sci. USA* **84**, 8360-8364.
12. Cowman, A.F., Morry, M.J., Biggs, B.A., Cross, G.A.M. & Foote, S.J. (1988) *Proc. Natl. Acad. Sci. USA* **85**, 9109-9113.
13. Peterson, D.S., Walliker, D. & Wellems, T.E. (1988) *Proc. Natl. Acad. Sci. USA* **85**, 9114-9118.
14. Snewin, V.A., England, S.M., Sims, P.F.G. & Hyde, J.E. (1989) *Gene* **76**, 41-52.
15. Zolg, J.W., Plitt, J.R., Chen, G. & Palmer, S. (1989) *Mol. Biochem. Parasitol.* **36**, 253-262.
16. Chen, G. and Zolg, J.W. (1987) *Mol. Pharmacol.* **32**, 723-730.
17. Matteucci, M. D. & Caruthers, M. H. (1980) *Tetrahedron Letters* **21**, 719.
18. Matteucci, M. D. & Caruthers, M. H. (1981) *J. Am. Chem. Soc.* **103**, 3185.
19. Beaucage, S.L. & Caruthers, M.H. (1981) *Tetrahedron Letters* **22**, 1859.
20. Banerjee, C.K., Bennet Jr., L.L., Brockman, W.R., Sani, B.P. & Temple Jr., C. (1982) *Anal. Biochem.* **121**, 275-280.

21. Dann, J.G., Ostler, G., Bjur, R.A., King, R.W., Scudder, P., Turner, P.C., Robert, G.C.K., & Burgen, A.S.V.(1976) *Biochem. J.* **157**, 559-571.
22. Futterman, S. (1957). *J. Biol. Chem.* **228**, 1031-1038.
23. Maniatis, T., Fritsch, E.F. & Sambrook, J. (1982) *Molecular cloning..* Cold Spring Harbor Laboratory, Cold Spring Harbor, NY.
24. Hanahan, D. (1983) *J. Mol. Biol.* **166**, 557-580.
25. Birnboin, H.C. & Doly, J. (1979) *Nucl. Acids Res.*, **7**, 1513-1525.
26. Weislander, L. (1979) *Anal. Biochem.* **98**, 305-309.
27. Grumont, R., Sirawaraporn, W. and Santi, D.V. (1988) *Biochemistry* **27**, 3776-3784.
28. Sanger, F., Nicklen, S. & Coulson, A. R. (1977) *Proc. Natl. Acad. Sci. U.S.A.* **74**, 5463-5467.
29. Meek, T. D., Garvey, E. P., & Santi, D. V. (1985) *Biochemistry* **24**, 678-686.
30. Read, S.M. & Northcote, D.H. (1981) *Anal. Biochem.* **116**, 53-64.
31. Laemmli, U.K. (1977) *Nature* **227**, 680-685.
32. Matsudaira, P. (1987) *J. Biol. Chem.* **261**, 10035-10038.
33. Santi, D.V., McHenry, C.S. and Sommer, H. (1974) *Biochemistry* **13**, 471-480.
34. Belfort, M., Moelleken, G., Maley, G.F., & Maley, F. (1983) *J. Biol. Chem.* **258**, 2045-2061.
35. Garvey, E.P. & Santi, D.V. (1985) *Proc. Natl. Acad. Sci. U.S.A.* **82**, 7188-7192.
36. Baker, T.A., Grossman, A.D. & Gross, C. A. (1984) *Proc. Natl. Acad. Sci. U.S.A.* **81**, 6779-6783.
37. Chen, G., Mueller, C., Wendlinger, M. & Zolg, J.W. (1987) *Mol. Pharmacol.* **31**, 430-437.
38. Garrett, C.E., Corderre, J.A., Meek, T.D., Garvey, E.P., Claman, D.M., Beverley, S.M. & Santi, D.V. (1984) *Mol. Biochem. Parasitol.* **11**, 257-265.

COAT PROTEIN MEDIATED PROTECTION AND THE POTENTIAL FOR ITS APPLICATION IN AGRICULTURE

Roger N. Beachy

Department of Biology , Washington University
One Brookings Dr., Campus Box 1137, St.Louis, MO 63130 USA

SUMMARY

With the development of techniques for the introduction of foreign genes into plant cells and the regeneration of transgenic plants, a number of different approaches have been taken to attempt to produce resistance to virus diseases. Those most commonly used have been (1) the expression of antisense RNAs; (2) the expression of satellite RNAs and defective interfering RNAs and DNAs; and (3) the expression of genes encoding viral capsid (coat) proteins. Of these the latter has been generally accepted because of its ready application to control a variety of virus diseases in a number of different crops. Because we recently published a review of the current state of the field (Beachy et al, 1990) this brief review will highlight the status of the field.

COAT PROTEIN MEDIATED RESISTANCE

What is it? In 1986 we reported that expression of a gene that instructs the transformed cell, and regenerated transgenic tobacco plants, to produce the coat protein (CP) of tobacco mosaic virus (TMV) confers resistance to infection by TMV (Powell-Abel et al., 1986). Plants that accumulate significant amounts of coat protein either escaped infection or, if infected, developed few if any symptoms, or developed mild disease symptoms after a significant delay in time. This technique, therefore, produced plants that are resistant to virus infection, and represents an example of the way that biotechnology through genetic engineering can make significant contributions to plant breeding.

Since the first publication of engineered resistance, which we have called "coat protein mediated resistance", there have been a growing number of examples of virus resistance through the use of this biotechnology. As summarized in the recently published review (Beachy et al, 1990) the list includes the following viruses, and plants:

Source of CP Gene	Transgenic Resistant Plant	Reference
Tobacco Mosaic Virus	Tobacco	Powell-Abel et al(1986)
	Tomato	Nelson et al (1987)

Biotechnology and Environmental Science: Molecular Approaches
Edited by S. Mongkolsuk et al., Plenum Press, New York, 1992

49

Source of CP Gene	Transgenic Resistant Plant	Reference
Tomato Mosaic Virus	Tomato	Sanders et al (submitted)
Alfalfa Mosaic Virus	Tobacco	Turner et al (1987)
	Tomato	Turner et al (1987)
	Alfalfa	Hill et al (submitted)
Cucumber Mosaic Virus	Tobacco	Cuozzo et al (1988)
	Cucumber	Gonsalves (unpublished)
Tobacco Streak Virus	Tobacco	vanDun et al (1988)
Potato Virus X	Tobacco	Hemenway et al (1988)
	Potato	Hoekema et al (1989)
		Kaniewski et al (1990)
Potato Virus Y	Tobacco	Stark et al (1989)
	Potato	Kaniewski et al (1990)
Tobacco Etch Virus	Tobacco	Stark et al (1989)
Tobacco Rattle Virus	Tobacco	vanDun et al (1988)
Potato Leafroll Virus	Potato	Kawchuk et al (1990)
Potato Virus S	Tobacco	MacKenzie et al (1990)

In each of the examples of CP-mediated resistance the transgenic plants harbor a gene that encodes the CP indicated providing resistance against the virus from which the gene was taken. In some cases, the gene gives resistance to viruses is closely related in structure to the virus from which the gene was taken.

GENERALIZED STRUCTURE OF A COAT PROTEIN GENE

Transcriptional Promoter	CP cDNA	3' end regulatory sequence

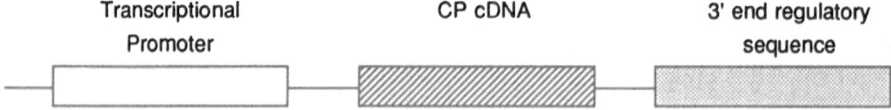

For example the CP gene of TMV gave resistance against the following tobamoviruses: tomato mosaic virus, pepper mild mottle virus, tobacco mild green mosaic virus, ondontoglossum ringspot virus, and ribgrass mosaic, but much less resistance against the distantly related tobamovirus sunn hemp mosaic virus (Nejidat and Beachy, 1989).

Similar results were obtained in a study with potyviruses (Stark and Beachy, 1989); that is, a CP gene from one potyvirus provided resistance against several different potyviruses. On the other hand, the TMV CP gene (or CP gene from other viruses) provided little or no resistance against viruses from other groups (Anderson et al, 1989). These reports document that coat protein mediated resistance provides a more broad type of disease resistance than many standard single genes for resistance, but does not "induce" a host response that causes resistance to all pathogens (Carr et al, 1989).

It is important to note that CP mediated resistance does not create plants that are "immune" to infection, and high levels of inoculum are, in some cases, able to overcome resistance. In other cases, however, plants are resistant to very high levels of inoculum, as in the cases of resistance against cucumber mosaic virus (Cuozzo et al, 1988), potato virus X (Hemenway et al, 1988) and potato virus Y and tobacco etch virus (Stark et al, 1989). In the case of resistance to TMV, plants that expressed the CP gene were susceptible to inoculum levels of 10 mg/ml, while plants that were not genetically modified succumbed to inoculum levels of 0.001-0.01 mg/ml

A recent study by Lawson et al (1990) demonstrated resistance against PVY that was vectored by aphids. These results provide evidence that CP mediated resistance can be effective under several types of inoculation regimes.

Mechanisms of disease resistance: At the current time we do not fully understand the cellular and molecular mechanisms of CP mediated resistance. However, the following statements can be made about resistance against TMV in tobacco and tomato, the two systems which have been studied most rigorously. Plants that are protected by coat protein mediated resistance:

1. are less likely to become infected than non-protected plants due to a reduction in the number sites of infection following inoculation than are non - resistant plants (Nelson et al, 1987). In practical terms this means that fewer plants become infected.

2. are less likely to develop systemic disease symptoms if infected than are plants that lack resistance. This means that fewer infected plants will develop symptoms that could reduce plant yield.

3. produce less virus, if infected, than do plants that lack resistance (Cuozzo et al, 1988; Stark and Beachy,1989). Thus, infected plants are less likely to serve as reservoirs for inoculum that might contribute to a disease epidemic.

These three characteristics, in concert, can contribute significantly to reducing disease epidemics caused by virus infections in agricultural fields. Not only are plants with coat protein mediated resistance less likely to become infected, those that become infected accumulate less virus, thus reducing plant-to-plant spread of the disease. This of course, serves to sustain crop yields.

While we do not know the precise mechanisms involved in coat protein mediated resistance, we strongly suspect that it involves interference with an event that is early in the infection cycle of the virus (Register and Beachy, 1988; 1989; Wu et al, 1990). In the TMV example it is suspected that resistance involves blocking the uncoating of the virus particle; therefore, the virus nucleic acid is not expressed, and the infection it is not initiated.

How does this mechanism affect the spread of virus once infection takes place? In some cases it may have little or no affect, as in the case with resistance against TMV (Wisniewski et all, 1990), while in other cases it may interfere with spread of the infection from cell to cell (Hemenway et al, 1989). While the virus infection may spread slowly within inoculated leaves, it may move very slowly from leaf to leaf. This effect is poorly understood, although it is known that the transcriptional promoter used to construct the coat protein gene can dramatically affect the level of disease resistance. (The transcriptional promoter controls the expression of the gene, it determines in what cell type, or at what stage in plant development, and/or at what level, the gene is expressed). Clark et al (1990) demonstrated that when the TMV CP gene is controlled by a promoter that causes expression in most or all cell types there is greater resistance than when the gene is expressed only in cells that contain chloroplasts. One can predict, then, that coat protein mediated resistance against specific classes of virus, such as those that are phloem-restricted, may require cell-specific or tissue specific expression of a coat protein gene to achieve high levels of resistance.

In conclusion, while we have gained considerable knowledge about coat protein mediated resistance there remains to be completed more and different research approaches to discern the precise cellular and molecular basis of resistance.

FIELD TRIALS WITH VIRUS RESISTANT PLANTS

There have been several field trials involving a limited number of types of transgenic plants that express different virus CP genes. It is, first, important to indicate that each of the tests performed have been preceded by approval granted from the U.S. Department of Agriculture and the Animal and Plant Health Inspection Service (APHIS) in the U.S.A. using guidelines proposed by a number of informed scientists. Limited release field trails of TMV resistant tomato plants clearly demonstrated resistance both against direct inoculation with TMV and incidental secondary inoculation that occurred during weeding of the plot. Furthermore, in this initial trial the resistant plant line was protected against yield losses from infection which reduced yield in the susceptible parental line by 25-30% (Nelson et al, 1988). Interestingly, although the TMV CP gene provided excellent resistance to tomato mosaic virus (ToMV, a tobamovirus related to TMV) in greenhouse studies (Nelson et al, 1988) there was only marginal protection under field conditions. On the other hand the ToMV CP gene gave excellent protection against ToMV (Kaniewski, W. and colleagues, Monsanto Company, St. Louis; unpublished data). Furthermore, plant lines that carry both the TMV and ToMV CP genes are resistant against both viruses (Kaniewski et al, unpublished data).

More recently there have been several small scale field tests of potato plants that contain CP genes of both potato virus X and potato virus Y. Whereas a double infection of non-resistant plants can lead to severe yield losses due to a synergistic reaction to a double infection, one of the transgenic plant lines was highly resistant to both viruses under field conditions (Kaniewski et al, 1990)

There have been other successful field trials, the results of which are unpublished, of other types of plants that also demonstrate the efficacy of coat protein mediated resistance. However, additional small scale, as well as widespread large field testing of genetically modified, virus resistant plants remain to be done. Nevertheless, there is high expectation that this new form of resistance will be a valuable asset to plant breeders as they search for new genes for resistance against virus diseases.

BIBLIOGRAPHY

Anderson E, Stark D, Nelson R, Tumer N, and Beachy R ((1989) Transgenic plants that express the coat protein gene of TMV or AlMV interfere with disease development of non-related viruses. Phytopathology 12:1284-1290.

Beachy R, Loesch-Fries S, Tumer N (1990) Coat protein-mediated resistance against virus infection. Annu. Rev. Phytopathol. 28:451-74.

Carr J, Beachy R, and Klessig D (1989) Are the PR1 proteins of tobacco involved in genetically engineered resistance to TMV? Virology 169:470-473.

Clark W, Register III J, Nejidat A, Eicholtz D, Sanders P, Fraley R, and Beachy R (1990) Tissue specific expression of the TMV coat protein in transgenic tobacco plants affect the level of coat protein mediated virus protection. Virology 179:640-647.

Cuozzo M, O' Connell K, Kaniewski W, Fang R, Chua X, Tumer N (1988) viral protection in transgenic tobacco plants expressing the cucumber mosaic virus coat protein or its antisense RNA. Bio/Technology 6:549-557.

Hemenway C, Fang R, Kaniewski W, Chua N, and Tumer N (1988) Analysis of the mechanism of protection in transgenic plants expressing the potato virus X coat protein or its antisense RNA. EMBO J. 7:1273-1280.

Hill K, Jarvis-Egan N, Halk E, Krahn K, Liao L, et al (1990) The development of virus-resistant alfalfa, Medicaco sativa L, Bio/Technology (submitted).

Hoekema A, Huisman M, Molendijk L, van den Elzen P, and Cornelissen B (1989) The genetic engineering of two commercial potato cultivars for resistance to potato virus X. Bio/Technology 7:237-278.

Kaniewski W, Lawson C, Sammons B, Haley L, Hart J, Delanny X, and Tumer N (1990) Field resistance of transgenic russet burbank potato to effect of infection by potato virus X and potato virus Y. Bio/Technology 8:750-754.

Kawchuk L, Martin R, McPherson J (1990) Resistance in transgenic plants expressing the potato leafroll luteovirus coat protein gene. Mol. Plant Microb. Interact. 3:301-307.

Lawson C, Kaniewski w, Haley L, Rozman R, Newell C, et al (1990) Engineering resistance to mixed virus infection in a commercial potato cultivar: resistance to potato virus X and potato virus Y in transgenic Russet Burbank. Bio/Technology 8:127-134.

MacKenzie D, Tremaine J (1990) Transgenic *Nicotiana debneyii* expressing viral coat protein are resistant to potato virus S infection. J. Gen. Virol 71:2167-2170.

Nejidat A., and Beachy R (1989) Low expression of TMV coat protein in transgenic tobacco plants under elevated temperature reduced plant resistance to TMV infection. Virology 173:531-538.

Nelson R, Powell-Abel P, and Beachy R (1987) Lesions and virus accumulation in inoculated transgenic tobacco plants expressing the coat protein gene of tobacco mosaic virus. Virology 158:126-132.

Powell-Abel P, Nelson R, De B, Hoffmann N, Rogers S, Fraley R, and Beachy R (1986) delay of disease development in transgenic plants that express the tobacco mosaic virus coat protein gene. Science 232:738-743.

Register III J, and Beachy R (1988) Resistance to TMV in transgenic plants results from interference with an early event in infection. Virology 166:524-532.

Register III J, Powell P, and Beachy R (1989) Genetically engineered cross protection against TMV interferes with initial infection and long distance spread of the virus. In: Molecular Biology of Plant-Pathogen Interactions. B. Staskowicz, P. Ahlquist, and O.C. Yoder eds., pp. 269-281. Alan R. Liss, New York.

Sanders P, B. Sanders, Kaniewski W, Haley L, La Vallee B, Delany X, Tumer N (1992) Field resistance of transgenic tomatoes expressing the tobacco mosaic virus or tomato mosaic virus coat proteingenes. Phytopathalogy (impress).

Stark D, and Beachy R (1989) protection against potyvirus infection in transgenic plants: Evidence for broad spectrum resistance. Bio/Technology 7:1257-1262.

Tumer N, O'Connell K, Nelson R, Sanders P, and Beachy R (1987) Expression of alfalfa mosaic virus coat protein gene confers cross-protection in transgenic tobacco and tomato plants. EMBO J. 6:1181-1188.

vanDun C, and Bol J (1988) Transgenic tobacco plants accumulating tobacco rattle virus coat protein resist infection with tobacco rattle virus and pea early browning virus. Virology 167:649-652.

Wu X, Beachy R, Wilson T, and Shaw J (1990) Inhibition of uncoating of tobacco mosaic virus particles in protoplasts from tobacco plants that express the viral coat protein gene. Virology 179:893-895.

MOLECULAR STRATEGIES IN THE INTERACTION BETWEEN
Agrobacterium AND ITS HOSTS

Eugene W. Nester* and Milton P. Gordon*#

*Department of Microbiology; # Department of Biochemistry
University of Washington, Seattle, WA

OVERALL FEATURES OF CROWN GALL TUMOR FORMATION

Agrobacterium tumefaciens induces a disease, crown gall, in a wide variety of dicotyledonous plants by transferring a piece of its tumor-inducing (Ti-) plasmid into the plant cell where it becomes integrated and functions in the plant. The overall features of this disease are illustrated in Fig. 1.

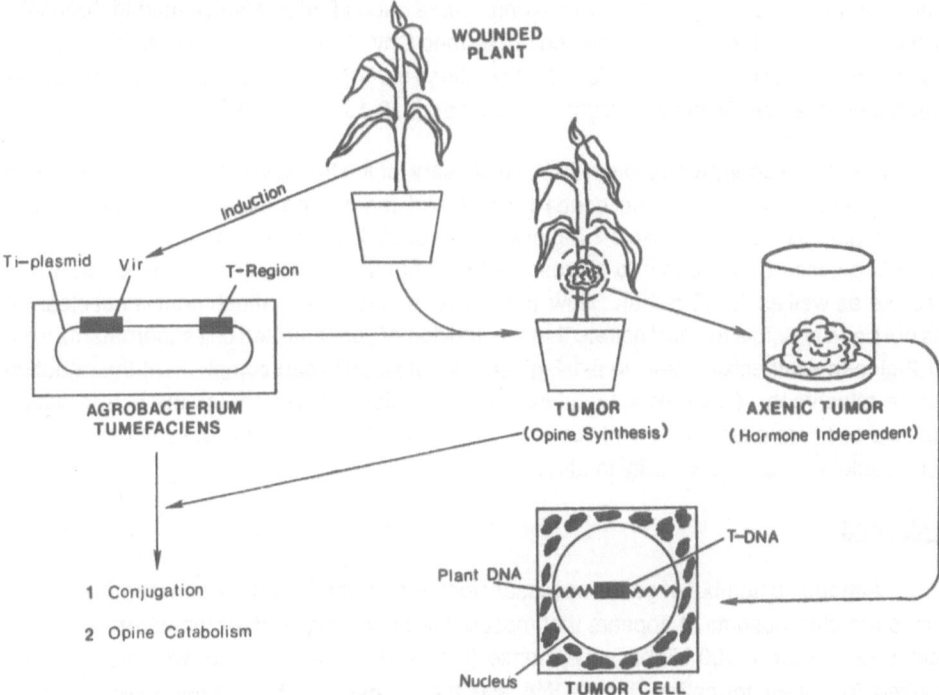

Fig.1 Overall features of crown gall tumor formation by *Agrobacterium*. It is possible to delete the T-DNA and insert useful genes under the control of plant promoters. The expression of these genes integrated into the plant DNA confers desired properties on the plant. This technology forms the basis of genetic engineering of plants using *Agrobacterium*.

Biotechnology and Environmental Science: Molecular Approaches
Edited by S. Mongkolsuk et al., Plenum Press, New York, 1992

The transferred and integrated DNA (T-DNA) codes for the synthesis of the two growth regulators, auxin and cytokinin, as well as for a group of amino acid derivatives termed opines. The expression of the genes for phytohormone synthesis, which are not subject to regulation by the plant, gives rise to the symptoms of crown gall tumor formation. The transfer of the T-DNA requires the expression of a variety of other genes on the Ti-plasmid, the virulence (*vir*) genes. These genes, which are involved in the processing and transfer of the T-DNA, are not expressed when Agrobacterium grows in the absence of plant cells, but are activated by plant cell metabolites synthesized by the wounded plant. As far as we are aware, the Agrobacterium-plant interaction is unique in that the end result is the transfer and integration of a piece of bacterial DNA into the plant chromosome. However, the interaction of this organism with the plant can serve as a model system for many other bacterial-plant interactions, since it involves such features as the attachment of the bacteria to production of phytohormones, auxin and cytokinin, as the virulence factors which result in the gall-like symptoms of the disease. All of these aspects can be found in other bacterial-plant interactions.

EARLY EVENTS IN THE TRANSFER OF T-DNA INTO PLANT CELLS

Attachment of bacteria to plant cells

Considerable evidence exists that *Agrobacterium* must bind to plant cells in order to cause crown gall tumors. However, the molecular basis of this attachment process remains elusive. A number of mutants of *Agrobacterium* have been identified which map to three loci, all of which map to the chromosome and are termed Chv. None of these mutants is capable of attaching and all are avirulent. Further, all of these mutants are involved with the synthesis or transport of a low molecular weight polysaccharide, β-1,2-glucan (Fig.2).

The *Chv*A codes for a protein that is necessary for the transport of the β-1,2-glucan into the periplasm; the *Chv*B gene codes for a 235kd membrane-associated protein which converts glucose into the cyclic β-1,2-glucan. The *exo*C locus codes for and enzyme which converts glucose 6-phosphate to glucose 1-phosphate, a step required for the synthesis of cellulose as well as β-1,2-glucan. How β-1,2-glucan functions in attachment is not clear. It has not been possible to demonstrate that the addition of concentrated cell supernatants from β-1,2-glucan synthesizing cells to β-1,2-glucan negative cells can complement the negative cells in attachment (Cangelosi, unpublished observation). Significantly, the β-1, 2-glucan negative cells are pleiotrophic and some property other than β-1,2-glucan synthesis may be responsible for the cells' inability to attach.

Ti-plasmid

Although a number of genes necessary for tumor formation have been identified which map to the chromosome, it appears that most of the genes required for tumor formation are located on the large 180 kilobase Ti-plasmid (Fig.3). In addition to the two sets of genes required for tumor formation, the T-DNA and the vir genes, other genes which play no significant role in tumor formation have also been identified. These include genes for the metabolism of the opines as well as housekeeping genes concerned with the replication of the plasmid.

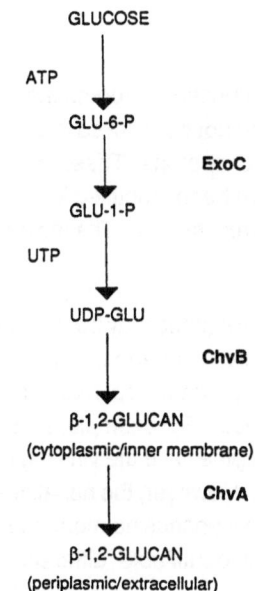

GLUCOSE

ATP

GLU-6-P

ExoC

GLU-1-P

UTP

UDP-GLU

ChvB

β-1,2-GLUCAN

(cytoplasmic/inner membrane)

ChvA

β-1,2-GLUCAN

(periplasmic/extracellular)

**BIOCHEMICAL ANALYSIS OF
β-1,2-GLUCAN MUTANTS**

Fig. 2 Identification of the site of mutations involving the synthesis and transport of β-1,2-glucan

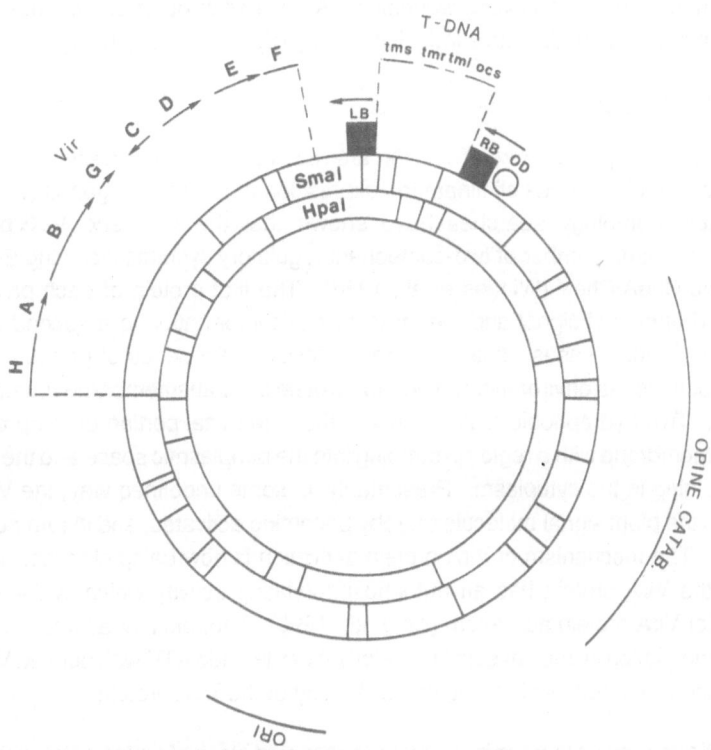

Fig. 3 Restriction enzyme map of T-plasmid indicating restriction sites produced by SmaI and HpaI.

Designation of _vir_ genes

The virulence genes are comprised of approximately 35 kilobases of DNA and are essential for tumor formation; they are not transferred into the plant. They can be divided into six highly studied transcription units, or operons. These include _vir_A, _vir_G, _vir_B, _vir_C, _vir_D, and _vir_E. Two additional _vir_ operons have been identified in the octopine Ti _vir_ regions (_vir_H and _vir_F), but they have not been studied extensively. The most relevant data on the _vir_ genes are summarized in Fig. 4.

It seems likely that additional _vir_ genes may be identified in the future. These probably will be recognized as genes which are required for tumor formation on some, but not other plants and/or as genes which are required for optimal tumor formation. Different strains of _Agrobacterium_ have different _vir_ genes. For example, strains of _Agrobacterium_ that induce tumors that synthesize the opine octopine have the _vir_H region, whereas strains which induce nopaline synthesizing tumors do not. However, the nopaline-inducing strains have a _vir_ gene coding for the synthesis of isopentenyl adenosine monophosphate at the same relative map position. There are other less well-identifiable differences between these two types of Ti-plasmids.

Chemical Inducer of _vir_ Genes

The induction of _vir_ genes depends upon the synthesis of plant signal molecules synthesized by the wounded plant. One such molecule which has been identified is acetosyringone (3,5 dimethoxy-4-hydroxy acetophenone) (Stachel et al., 1985). This compound, which seems to be a derivative of a precursor of lignin biosynthesis, has been found in a wide variety of plants following wounding. A number of other compounds chemically related to acetosyringone also have inducing activity (Spencer et al., 1988).

Activation of VirA Protein

Two of the _vir_ genes, _vir_A and _vir_G, are required for the expression of all vir genes since mutations in either locus eliminate the expression of all other vir genes (Winans et al., 1986). Protein homology searches have shown that the VirA and VirG proteins are homologous to a large number of two-component regulatory systems including EnvZ/ompR, NtrB/NtrC and CheA/CheY (Winans et al., 1986). The first protein of each pair detects a particular environmental signal and then transfers this information to a second component which in turn, in most cases, activates the expression of a series of genes whose gene products respond to the environment. The VirA protein is a transmembrane protein (Winans, et al., 1989). Two hydrophobic regions anchor the N-terminal portion of the protein to the cytoplasmic membrane with a region protruding into the periplasmic space and the C-terminal domain remaining in the cytoplasm. Presumably, in some undefined way, the VirA protein interacts with the plant signal molecule thereby becoming activated, and in turn activates the VirG protein. The mechanism of VirA protein activation is now being clarified. It has been shown that the VirA protein has an autophosphorylating activity which is the most likely mechanism for VirA protein activation (Jin et al., 1990). The role that acetosyringone plays in autophosphorylation is unclear since the addition of γ-labeled ATP with purified VirA protein but without acetosyringone still results in the labeling of the VirA protein.

A histidine residue located in the highly conserved block of amino acids which is found in all of the VirA homologues is the amino acid phosphorylated. If this histidine is mutated to

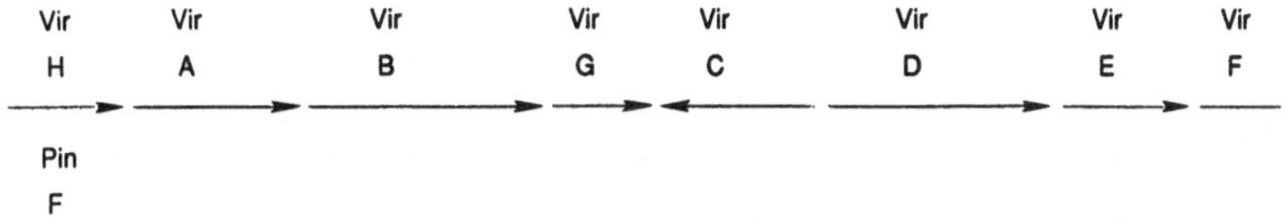

Vir H   Vir A   Vir B   Vir G   Vir C   Vir D   Vir E   Vir F

——→ ————→ —————→ ————→ ←———— —————→ ———→ ——

Pin F

Vir	Inducibility	Size (Kb)	ORF's	Function
A	+	2.8	1	Plant Signal Sensor
G	+	1.0	1	Transcriptional Activator
B	+	9.5	11	Pore?
D	+	4.5	4	Processing of T-DNA-Endonuclease
C	+	1.5	2	Processing of T-DNA
E	+	2.2	2	Single-Strand DNA Binding Protein
H	+	3.4	2	Cytochrome P450 Enzyme
F	?	?	?	?

Fig.4 *Vir* gene order and relevant information.

a glutamine residue which cannot be phosphorylated, then the *vir* genes are not inducible nor is the strain capable of inducing crown gall tumors. This demonstrates that phosphorylation of the VirA protein is essential for its proper functioning.

Interaction of VirA and VirG

The phosphate residue of the phosphorylated histidine in the VirA protein is transferred directly to the VirG protein (Jin, unpublished observation). Thus, when the phosphorylated VirA protein was mixed with unphosphorylated VirG protein and samples withdrawn at various times with increasing time of incubation, the phospho-VirA signal got weaker while the phospho-VirG signal got stronger. The phosphate transfer occured very rapidly five seconds after VirA and VirG proteins were mixed together transfer could be observed.

The phosphate bond in the VirG protein is highly unstable to both acid and base, which suggests that either an aspartic acid or glutamic acid is being phosphorylated. By appropriate techniques, the unstable phosphate-amino acid bond was stabilized and the protein was then cleaved and the cleavage products sequenced. From these studies, we conclude that a specific aspartic acid residue is phosphorylated in the VirG protein (Jin et al., unpublished observation). When this particular aspartic acid was mutated *in vitro* to asparagine by site-directed mutagenesis, the mutant was no longer able to induce *vir* genes nor was it able to induce tumors on kalanchoe leaves. This indicates that phosphorylation of the VirG protein is essential for its biological function.

Interaction of the VirG Protein with the "*vir* Box."

The VirG protein recognizes a 12 base pair conserved sequence called the "*vir* box" which is located upstream of each of the *vir* genes (Winans et al., 1987). Footprinting analysis of each of the promoter regions indicates that the VirG protein covers the "*vir* box" on both strands of the DNA (Jin et al., 1990). However, phosphorylation does not seem to be required for the binding of the VirG protein to the "*vir* box" region. The VirG protein was isolated from *E.coli* cells into which the gene had been cloned, and alkaline phosphatase was added to remove any phosphate from the protein. This preparation of the VirG protein still bound specifically to a region of the DNA that included the "*vir* box". However, these data do not rule out the possibility that the phosphorylated VirG protein may have a higher affinity for this region than the nonphosphorylated form. Alternately, the action of the phosphorylated VirG protein may be beyond the binding step such as the activation of transcription by RNA polymerase.

The overall features of the role of the VirA and VirG interaction with one another and the interaction of the VirG protein with the "*vir* box" are shown in Fig.5.

Role of Sugars in *vir* Gene Induction.

It is now clear that in addition to acetosyringone, sugars are also involved in *vir* gene induction in some way. This conclusion is based on the observation that a mutation in a particular locus on the chromosome reduced induction by acetosyringone significantly and thereby attenuated virulence (Huang et al., 1990). When this locus was cloned and sequenced, it was shown to code for a protein which is homologous to a glucose/galactose binding protein in *E. coli*. Recent experiments have demonstrated that a variety of monosaccharides, all components of the plant cell wall, are able to induce *vir* genes most dramatically in the presence of low levels of acetosyringone (Cangelosi, Ankenbauer, and Nester, unpublished observation).

VIR GENE PRODUCTS

Following activation of the *vir* genes, the T-DNA is processed for transfer into plant cells. A number of steps have been identified. One of the early stages in T-DNA processing involves the conversion of the supercoiled Ti-plasmid into a relaxed form by a topoisomerase, the product of the VirD1 protein (Ghai et al., 1989). The next step is a site-specific cleavage in the two 25 base pair direct repeats which bound the T-DNA, the right and left borders. This is accomplished by an endonuclease, the product of the VirD2 protein (Yanofsky et al., 1986). Considerable evidence exists that the VirD2 protein cleaves at virtually identical sites in the bottom strand of the right and left border sequences, thereby resulting in the formation of a single stranded intermediate. Following cleavage, the VirD2 protein remains covalently bound to the 5' end of the T-DNA (Young et al., 1988). The *vir*E2 gene codes for a single stranded DNA binding protein which associates with the T-DNA and probably protects this DNA in both the bacteria and the plant cell (Christie et al., 1988). The transfer intermediate can be looked upon as a single stranded DNA molecule attached to the VirD2 gene product at the 5' end, and coated by a single stranded DNA binding protein. There is some speculation that the VirD2 protein may also serve to target the DNA to the plant cell nucleus. The early steps in T-DNA processing are illustrated in Fig.6.

The exit of the T-DNA from the bacterial cell probably occurs through pores in the bacterial cell envelope formed by the gene products coded by the virB operon (11 open reading frames) (Ward et al., 1988). Many of these gene products have signal sequences and are associated with the bacterial cell envelope. Of considerable interest is the fact that the last open reading frame in this operon (number 11) is phosphorylated and has ATPase activity (Christie et al., 1989). Further, this protein is very similar to one required for the development of competence in the *Bacillus subtilis* DNA transformation system (Albano et al., 1989).

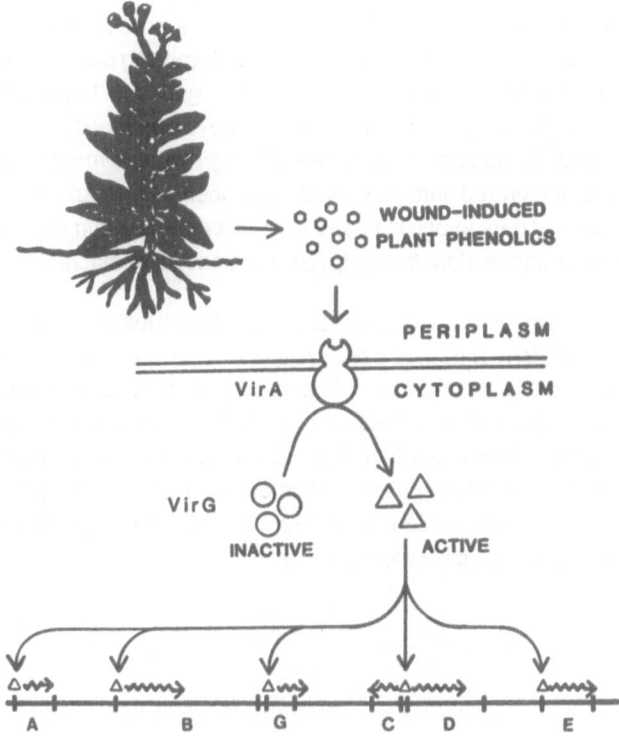

Fig.5 Model for the interaction of the VirA and VirG proteins with each other and with the "*vir* box."

From these studies, we envision that *Agrobacterium* conjugates with plant cells and transfers single stranded DNA from the bacterium to the plant cell. Once inside the plant cell, the DNA presumably recombines by a nonhomologous mechanism into a variety of sites on the plant chromosome. Once integrated, the T-DNA genes behave and function as plant genes.

EXPRESSION OF THE T-DNA

The T-DNA codes for a number of different transcripts, some of which have been functionally identified. Two of the best-characterized transcripts code for two enzymes of auxin synthesis and another codes for an enzyme of cytokinin synthesis (Binns et al., 1988). There is considerable evidence that these three genes are highly homologous to their counterparts in *P. savastanoi* (Yamada et al., 1985). These latter studies, carried out in Tsune Kosuge's laboratory, strongly suggest that the structural genes of the T-DNA are probably of bacterial origin and have not been "captured" from plants by bacteria. Other well-characterized genes of the T-DNA code for enzymes of opine synthesis.

SUMMARY

The formation of hyperplasias termed "galls" produced by *P.savastanoi* and the formation of crown gall tumors by *Agrobacterium tumefaciens* have a number of features in common. In both cases, the unregulated production of IAA and cytokinin at the site of infection is required for gall formation. Thus, both compounds, the product of bacterial genes, are virulence factors in both diseases, and therefore, it is not surprising that the disease symptoms are similar. Further, both virulence factors are coded by genes on large plasmids and have a remarkable degree of homology with each other. However, in *Pseudomonas* the genes are constitutively expressed in the bacterium, whereas in *Agrobacterium* expression occurs only after the genes are integrated into the plant chromosome. Apparently, the structural component of the gene is prokaryotic in nature whereas the upstream regulatory sequences are eukaryotic and respond to plant transcription and translation machinery.

Disease production by *P. savastanoi* appears to be a much simpler process than that by *Agrobacterium*. In the former case, the two phytohormones are synthesized by the bacteria at the site of infection and living bacteria are necessary for continued symptomotology. On the other hand, *Agrobacterium* has evolved a very complex mechanism for introducing a piece of its plasmid into plant cells where it functions. Consequently, the continued growth of the bacteria at the wound site is not necessary for tumor formation. The vir genes of *Agrobacterium* apparently do not have a counterpart in *Pseudomonas*, although its virulence plasmid has not been subjected to extensive mutagenesis.

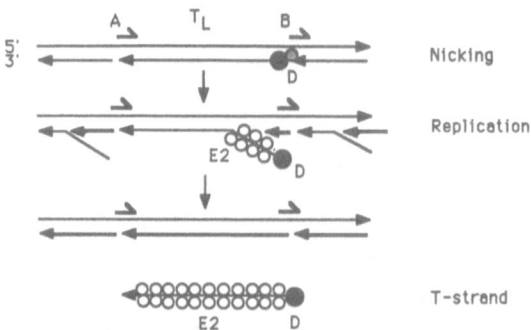

Fig.6 Proposed model for formation of T strand intermediate in transfer process. A and B refer to the left and right borders, respectively. T_L refers to the leftward T-DNA. The E2 is a single strand DNA binding protein and is a product of the *vir* E2 gene. The D refers to the product of the *vir* D2 locus which binds tightly to the 5' end of the T strand.

It is also instructive to compare the reasons why these two organisms induce galls. The most reasonable explanation as to why *Agrobacterium* induces tumors is that the T-DNA codes for the synthesis of opines which can serve as a source of carbon, nitrogen, and energy for the inducing strain of *Agrobacterium* but very few other organisms in the soil environment (Tempe et al., 1982). These opines further promote the transfer of the T-DNA from plasmid-containing to non-plasmid containing strains. It has been postulated that *Pseudomonas* forms galls in order to provide a protected environment for the bacteria during the hot, dry summers characteristic of regions where its hosts, olive and oleander, flourish. Only in this protected environment do the bacteria grow to high concentrations.

ACKNOWLEDGMENTS

We would like to thank the members of our research groups both past and present for providing much of the data presented in this paper. This work was supported by NIH grant 5RO1GM32618, NSF #DMB8704292, and USDA #88-37234-3618.

REFERENCES

Albano, M., Breitling, R., and Dubnau, D.,1989, Nucleotide sequence and genetic organization of the *Bacillus subtilis* comG operon. *J. Bacteriol.*, **171**:5286-5404.

Binns,A., and Thomashow, M., 1988, Cell biology of *Agrobacterium* infection and transformation of plants. *Ann. Rev. Microbiol.* **42**:575-606.

Christie, P.,et al., 1988, The *Agrobacterium tumefaciens vir*E2 gene product is a single-stranded-DNA-binding protein that associates with T-DNA. *J.Bacteriol.* **170**:2659-2667.

Christie, P., et al., 1989, A gene required for transfer of T-DNA to plants encodes an ATPase with autophosphorylating activity. Proc. Natl. Acad.Sci. USA, **86**:9677-9681.

Ghai, J., and Das, A., 1989, The VirD operon of Agrobacterium tumefaciens Ti-plasmid encodes a DNA-relaxing enzyme. *Proc. Nat. Acad. Sci.* **89**:3109-3113.

Hung, M-LW., et al.,1990, A chromosomal Agrobacterium tumefaciens gene required for effective plant signal transduction. J.Bacteriol. **172**:1814-1822.

Jin,S.,et al., 1990, The VirA protein of *Agrobacterium tumefaciens* is autophosphorylated and is essential for vir gene regulation. *J. Bacteriol.* **172**:525-530.

Jin,S., et al., 1990, The regulatory VirG protein specifically binds to a cis-acting regulatory sequence involved in transcriptional activation of *Agrobacterium tumefaciens* virulence genes. *J. Bacteriol.* **172**:531-537.

Spencer, P.and Towers, G.H.N., 1988, Specificity of signal compounds detected by *Agrobacterium tumefaciens. Phytochemistry* **27**:2781-2785.

Stachel, S., et al., 1985, Identification of the signal molecules produced by wounded plant cells that activate T-DNA transfer in *Agrobacterium tumefaciens*. Nature (London) **318**: 624-629.

Ward, J., et al., 1988, Characterization of the virB operon from an *Agrobacterium tumefaciens* Ti-plasmid. *J. Biol. Chem.* **263**:5804-5814.

Winans, S., et al., 1987, The role of virulence regulatory loci in determining *Agrobacterium* host range, **In:***Plant Molecular Biology* (D.Von Wettstein and N.Chua, eds), Plenum Publishing Corp., New York, pp. 573-582.

Winans, S.,et al., 1989, A protein required for transcriptional regulation of *Agrobacterium* virulence genes spans the cytoplasmic membrane. *J.Bacteriol.* **171**:1616-1612.

Yamada, T., et al., 1985, Nucleotide sequences of the *Pseudomonas savastanoi* indole acetic acid pathway genes show homology with *Agrobacterium tumefaciens* T-DNA. *Proc. Natl. Acad. Sci. USA* **82**:6522-6526.

Yanofsky, M., et al., 1986, The virD operon of *Agrobacterium tumefaciens* encodes a site-specific endonuclease. *Cell* **417**:471-477.

Young, C., and Nester E., 1988, Association of the VirD2 protein with the 5' end of T strands in *Agrobacterium tumefaciens. J. Bacteriol.* **170**:3367-3374.

STRUCTURE OF PEROXIDASE ISOZYME GENES IN BRASSICA AND THEIR EXPRESSION

A. Shinmyo, K. Fujiyama,* H. Takemura, N. Okada, and
M. Takano

Department of Fermentation Technology, and
*International Center of Cooperative Research in Biotechnology, Osaka
University, Suita, Osaka 565, Japan

SUMMARY

We have isolated 5 genomic DNAs (prxC1a, prxC1b, prxC1c, prxC2 and prxC3) coding for horseradish peroxidase (HRP). All 5 genes consisted of 4 exons and 3 introns, and the positions of introns in the coding regions were same in all genes. Putative promoter sequences and polyadenylation signals were also found. Two peroxidase genes (prxCa and prxEa) of Arabidopsis thaliana, which belongs to the same family of horseradish, Brassica,were also cloned. The structures and sequences of the coding regions of prxCa and prxEa were very similar to prxC1b and prxC3, respectively.

Transcripts of all HRP genes were abundant in horseradish cultured cells, but organ-specific expression was observed in horseradish plant. prxC1 was expressed in root and stem, C2 in stem, and C3 in root.

5'-Noncoding regions of HRP isozyme genes were joined to β-glucuronidase (GUS) reporter gene and introduced into tobacco protoplast by electroporation. Promoter activity was detected in every construct, especially prxC2 gene showed several times higher activity of CaMV35S promoter. The promoter-GUS fused genes were transferred to tobacco leaf disk using binary vector system with Agrobacterium tumefaciens. Strong GUS expression was observed in root of transgenic tobacco, but not in leaf. When callus was induced from leaf, GUS activity enhanced 15 -50 time of that in leaf in all promoter regions as observed in horseradish callus.

1. INTRODUCTION

Plant peroxidase (EC 1. 11. 1. 7 donor: hydrogen peroxide oxidoreductase) is a glycoprotein containing protoheme IX and consists of many isozymes and/or isoforms. Several physiological functions of peroxidase in plant have been suggested, such as removal of H_2O_2, oxidation of toxic reductants, biosynthesis and degradation of lignin, defence response toward wounding, and metabolism of auxin. These functions might be dependent on each isozyme/

Biotechnology and Environmental Science: Molecular Approaches
Edited by S. Mongkolsuk et al., Plenum Press, New York, 1992

65

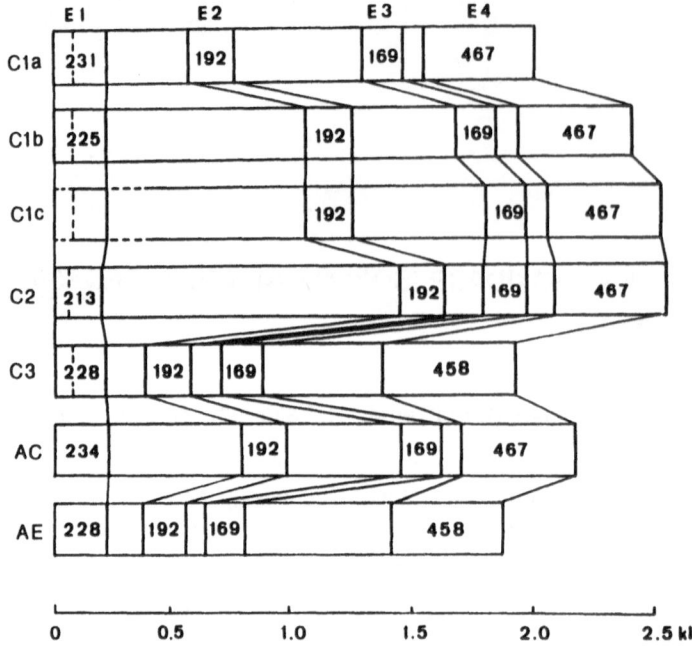

Fig. 1. The structure of the genomic DNA coding for <u>Brassica</u> peroxidase.
C1a - C3: HRP isozyme genes, AC and AE: <u>Arabidopsis</u> isozyme
genes. Numbers indicate the length of each exon (E1 - E4)
in bp. Vertical broken lines show the position corresponding
to N-terminus of the mature enzymes.

isoform in each plant tissue. Characterization of the gene structure and mode of gene
expression of peroxidase isozymes is an important approach to understand their biological
functions. Lagrimini <u>et al</u>. (1) constructed transgenic tobacco which was introduced tobacco
anionic peroxidase cDNA. Transformed tobacco had the unique phenotype of chronic severe
wilting through loss of turgor in leaves at the time of flowering.

Horseradish (<u>Armoracia rusticana</u>) is an important source of peroxidase which is used
in enzyme immunoassays and diagnostic assays, and contains almost 30 isozymes/isoforms
(2). Extensive studies have been done on the structure and function of horseradish peroxidase
(HRP). Here, we describe the structure of HRP isozyme genes and their expression. Structure
of peroxidase genes in <u>Arabidopsis thaliana</u>, which belongs to the same family, <u>Brassica</u>, to
horseradish, are also shown.

2. CLONING OF HRP GENES

We have isolated 3 cDNA clones (pSK1, pSK2 and pSK3) of HRP neutral isozymes
based on the amino acid sequence established by Welinder (3), and 5 genomic DNA
(<u>prxC1a</u>, <u>prxC1b</u>, <u>prxC1c</u> (partial), <u>prxC2</u>, and <u>prxC3</u>) from the gene library constructed from
horseradish cultured cells (4,5). <u>prxC1a</u>, <u>prxC1b</u> and <u>prxC1c</u> corresponded to cDNA inserts in
pSK1, pSK2, and pSK3, respectively.

These 5 genes were predicted to consist of 4 exons and 3 introns. All introns were
AT-rich and had consensus GT and AG at 5' and 3' termini, respectively. The lengths of exons
2 and 3 (192 bp and 169 bp, respectively) were the same in the <u>prx</u> genes (Fig. 1). The

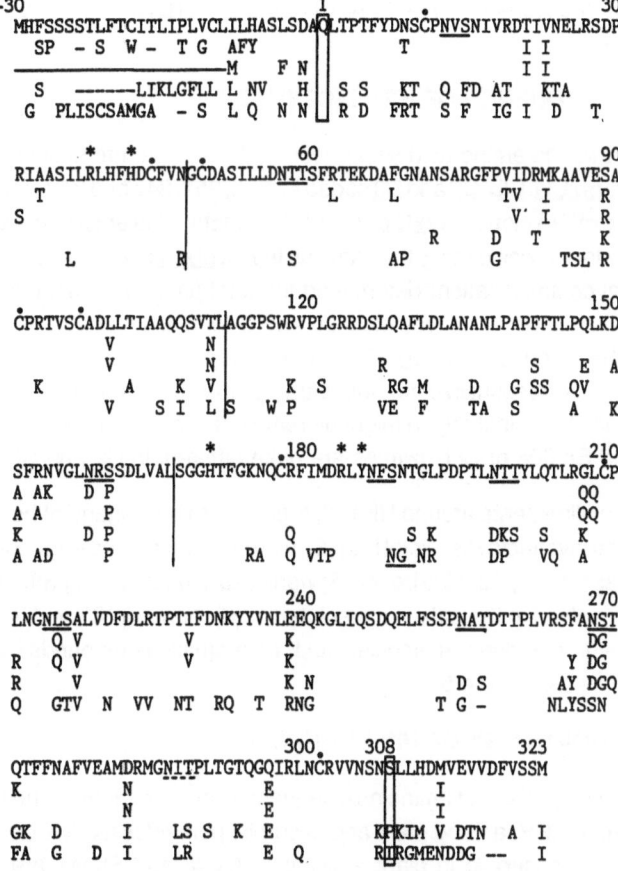

```
      -30                        1                     30
Cla  MHFSSSSTLFTCITLIPLVCLILHASLSDAQLTPTFYDNSCPNVSNIVRDTIVNELRSDP
Clb   SP - S  W- T G AFY                T              I I
Clc  ----------------M   F N                           I I
C2    S   ------LIKLGFLL L NV   H    S S    KT Q FD AT   KTA
C3    G   PLISCSAMGA - S L Q    N N    R D    FRT  S F  IG I  D  T

      *  *                       60                    90
     RIAASILRLHFHDCFVNGCDASILLDNTTSFRTEKDAFGNANSARGFPVIDRMKAAVESA
       T                          L       L            TV I       R
       S                                               V' I       R
                                      R       D    T          K
         L                 R          S         AP       G      TSL R

                                 120                   150
     CPRTVSCADLLTIAAQQSVTLAGGPSWRVPLGRRDSLQAFLDLANANLPAPFFTLPQLKD
        V              N                    R                S  E  A
        V              N                                     Q V
      K        A   K   V          K S       RG M     D    G SS QV
        V      S I L  S        W P          VE F     TA   S        A   K

                 *       180 **                        210
     SFRNVGLNRSSDLVALSGGHTFGKNQCRFIMDRLYNFSNTGLPDPTLNTTYLQTLRGLCP
      A AK   D P                                                   QQ
      A A     P                                                    QQ
      K      D P                  Q           S K       DKS S  K
      A AD    P                  RA  Q VTP     NG NR     DP    VQ A

                                 240                   270
     LNGNLSALVDFDLRTPTIFDNKYYVNLEEQKGLIQSDQELFSSPNATDTIPLVRSFANST
        Q V          V           K                             DG
      R Q V          V           K                           Y DG
      R   V                      K N              D S         AY DGQ
      Q   GTV  N  VV NT RQ T     RNG             T G -        NLYSSN

                                 300   308      323
     QTFFNAFVEAMDRMGNITPLTGTQGQIRLNCRVVNSNSLLHDMVEVVDFVSSM
     K              N          E             I
                    N          E             I
     GK D      I   LS S  K    E             KPKIM V DTN A  I
     FA G    D I   LR     E   Q             RIRGMENDDG --    I
```

Fig. 2 Amino acid sequences of five HRP isozymes deduced from nucleotide sequences. Those amino acids different from Cla isozyme are shown in Clb, Clc, C2 and C3 isozymes. Putative N- and C-termini are boxed, and the positions interrupted by introns on DNA sequence are shown by vertical bars. Distal His42, proximal His 170, and other hypothetical functional residues are indicated by asterisks. Cys residues are also shown by heavy dots. Putative N-glycosylation sites, Asn-X-Ser/Thr, are underlined.

homology of the nucleotide sequences of the coding regions was 90 - 93% in the prxCl family, 71% between prxCla and prxC2, 66% between prxCla and prxC3, and 63% between prxC2 and prxC3.

A putative promoter sequence, the TATA box, and the upstream consensus promoter in eukaryotes, the CAAT box, were found in every gene. A possible polyadenylation signal, AATAAA, was also found downstream of the TGA termination codon.

Two peroxidase genes (prxCa and prxEa) of Arabidopsis thaliana were obtained from a genomic library using pSKl as a probe. Both genes were also concluded to consist of 4 exons from the comparison with HRP genes, and putative exons 2 and 3 were same length to those in HRP genes. The homology of the nucleotide sequences of the coding regions was 65%

between prxCa and prxEa, but the value was 91% between prxCa and prxClb, and 89% between prxEa abd prxC3.

3. AMINO ACID SEQUENCE OF HRP ISOZYMES

Fig. 2 shows the amino acid sequences deduced from the nucleotide sequences of 5 HRP isozymes. prxCla, Clb, C2 and C3 coded for 353, 351, 347 and 349 amino acid residues, respectively. The cDNA clone of prxClc was not full length. The amino acid sequence of HRP C found by Welinder (3) was exactly the same as that of prxCla from Glnl to Ser308. Recently, Mikami (personal communication) determined the total amino acid sequence of HRP basic isozyme E, in which N-terminus was blocked and C-terminus was Arg306, and it was 88% homologous to that deduced from prxC3. The extrapeptide of 24-30 amino acid residues at N-terminus might be responsible for the intra-cellular localization or secretion, and the extrapeptide in the C-terminal region might be removed to form mature HRP. The C-terminal sequence around Ser308 of CI isozymes are quite different in C2 and C3.

Sequence alignments around His170, predicted to be a ligand of heme iron (proximal His), and His42, an invariable His (distal His) (6), were well conserved in the 5 isozyme proteins. Eight Cys residues forming disulfide bonds (3) were also conserved. Arg38, Phe41 and Phe45, proposed to be involved in the catalytic mechanism (6), and Arg183 and Tyr185, suggested to interact with hydrogen donor molecules such as aromatic compounds by Sakurada et al. (7) were also conserved in the 5 isozymes.

4. SEQUENCE HOMOLOGY OF HIS REGION

Fig. 3 is a comparison of amino acid sequences in the regions of the two His residues, His42 and His170 in HRP, among plant and microbial peroxidases. It is clear that sequence alignment is highly conserved in plant enzymes. Arg38 and Phe41 are conserved in all peroxidase, although Phe41 is replaced by another aromatic amino acid, Trp, in yeast (8) and bacterial peroxidse (9, 10). Phe45 is conserved only in plant enzymes. Sequence alignment in His regions was also conserved in Arabidopsis peroxidases (data not shown).

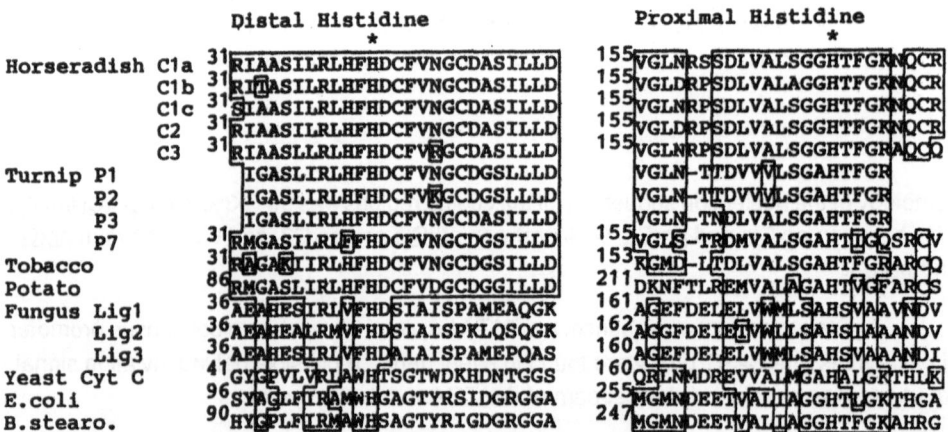

Fig. 3 The amino acid sequence comparison of proximal and distal histidine regions of plant and microbial peroxidase. Conserved amino acid substitutions are boxed. Asterisk indicate proximal and distal histidine residues. Numbers stand for those from N-terminal of first residues on each lines.

Peroxidase of E. coli (8) and B. stearothermophilus (9) are larger proteins with activities of peroxidase and catalase, and are less homologous to HRP. Yeast cytochrome c peroxidase (8) and fungal lignin peroxidase (11,12) have similar molecular masses to HRP, but homologies to HRP were less than 30%.

5. ORGAN-SPECIFIC EXPRESSION OF HRP GENES

The nucleotide sequences of 5' and 3' flanking regions of HRP genes suggested that all 5 gene is functional gene. Northern blot analysis was carried out using the 4th exon of prxCla, prxC2, and prxC3 as a probe, since the exon was not homologous in each other. Abundant transcript of all gene was detected in horseradish cultured cells. Organ-specific expression was observed in horseradish plant. prxCl was expressed in root and stem, C2 in stem, and C3 in root. Root-specific expression of Arabidopsis genes was also observed.

6. ACTIVITY OF HRP PROMOTER IN TOBACCO

About 500 bp upstream region of each HRP gene was ligated to β-glucuronidase (GUS) structure gene, and the chimeric genes were introduced to tobacco (Nicotiana tabacum BY2) protoplast by electroporation. Transient expression of GUS gene was detected in all construct, especially prxC2 promoter showed the activity of 4-5 times of CaMV35S promoter.

These constructs were introduced to tobacco leaf disk together with kanamycin-resistant gene using the Ti binary vector/Agrobacterium tumefacience system. GUS activity in Km^r-transgenic tobacco containing prxCla, prxClb, and prxC2 promoter was highest in root. prxC2 promoter activity was also observed in stem and reaf. When callus was induced from transgenic tobacco leaf, GUS activity increased 15 - 50 times under the control of these promoters as in horseradish cultured cells.

REFERENCES

1. Lagrimini, L. M., Bradford, S. and Rothstein, S.: The Plant Cell, 2,7 -18 (1990).
2. Hoyle, M. C.: Plant Physiol., 60, 787 - 793 (1977).
3. Welinder, K.G.: FEBS Lett., 72, 19 - 23 (1976).
4. Fujiyama, K., Takemura, H., Shibayama, S., Kobayashi, K., Choi, J. -K., Shinmyo, A., Takano, M., Yamada, Y. and Okada, H.: Eur. J. Biochem., 173, 681 - 687 (1988).
5. Fujiyama, K., Takemura, H., Shimyo, A., Okada, H. and Takano, M.: Gene, 89, 163 - 169 (1990).
6. Welinder, K. G.: Eur. J. Biochem., 151, 497 - 503 (1985).
7. Sadurada, J., Takahashi, S. and Hosoya, T.: J. Biol. Chem., 261, 9567-9662 (1986).
8. Kaput, J., Goltz, S. and Blobel, G.: J. Biol. Chem., 257, 15054 - 15058 (1982).
9. Triggs-Raine, B. L., Doble, B. W., Mulvey, M.R., Sorby, P.A. and Lowen, P.C.: J. Bacteriol., 170, 4415-4419 (1988).
10. Loprasert, S., Negoro, S., Okada, H.: J. Bacteriol. 171, 4871 - 4875 (1989).
11. Tien, M. and Tu, C.-P. D.: Nature, 326, 520-523 (1987).
12. de Boer, H. A., Zhang, H.A., Collins, C. and Reddy, C. A.: Gene, 60, 93 - 102 (1987).

6. ORGAN-SPECIFIC EXPRESSION OF HRP GENES

GREEN REVOLUTION AND GENE REVOLUTION

S.D. Kung

Center of Agricultural Biotechnology, Maryland Biotechnology Institute and
Department of Botany, The University of Maryland System
College Park

I. INTRODUCTION

"Food is the haven for the people, agriculture is the foundation of the country," this ancient Chinese motto expresses truthfully the importance of agriculture as our ancestors moved from hunting and gathering to cultivating and improving crops for their primary sustenance. Since all food is derived directly or indirectly from plants, selection and upgrading of plant products constituted the first step toward modern plant breeding. plant breeding became an important branch of modern science immediately after the rediscovery of Mendel's laws of genetics (Borlaug 1983).

It is with pride that we mention that the conventional plant breeding techniques have resulted in dramatic crop improvements and will continue to provide future benefit. Among the 3,000 species of plants used for food, only 29 species are major crops. These species include eight cereals, seven legumes, seven oil seeds, three root crops, two sugar crops and two tree crops (Borlaug 1983). Collectively, these provide the proteins, calories, vitamins and minerals necessary for the human diet. However, there are strong pressures for further improvements in crop quality and quantity exerted from population growth, social demands, health requirements, environmental stresses and ecological considerations. World population is predicted to reach eight billion by the year 2010. To feed three billion additional people in the next twenty years will require dramatic increases in crop production, a formidable task by any standard. Conventional plant breeders and related scientists have worked diligently and skillfully to upgrade quality and raise yields by employing various crop improvement techniques with commendable results. For example, the combined production of 17 major crops in the U.S. increased over 242 percent between 1940 and 1980, while acreage under cultivation increased approximately 3 percent (Borlaug 1983). Had crop yields remained at 1940 levels until 1980, 177 million additional acres of good U.S. cropland would have been required to produce the 1980 harvest. Notwithstanding these impressive gains in productivity, there are limitations to conventional plant breeding technology, which have stimulated the development of more advanced technologies.

This author wishes to describe the road we have traveled from the era of the Green Revolution to the Gene Revolution. This transition was a response to the limitations of

Biotechnology and Environmental Science: Molecular Approaches
Edited by S. Mongkolsuk et al., Plenum Press, New York, 1992

71

conventional breeding techniques. In order to produce hybrid plants with much wider ranges of genetic variabilities, somatic hybrid plants to be of immediate and economic value, in turn, stimulated the development of transgenic plant engineering. These remarks focus on the leap from the green revolution to the gene revolution viewed from the perspective of a plant molecular biologist.

II. PLANT BREEDING, HYBRID PLANTS AND GREEN REVOLUTION

1. Plant Breeding

Mendel's laws of genetics provided the scientific basis for plant breeding for almost a century. Plant breeding principles consist of identifying and selecting desirable traits and combining these into one individual plant. Since all traits are controlled by genes located on chromosomes, plant breeding can be considered as the manipulation of chromosomes. In general, there are four major ways to manipulate plant chromosomes. First, similar chromosomes can be sorted out and retained in one individual plant to reach a homozygous state, a method termed pureline selection. Second, different chromosomes can be combined together to obtain a heterozygous state. This method is termed hybridization. Third, new genetic variability can be introduced through spontaneous or artificially-induced mutations. Finally, polyploidy also contributes to crop improvement. They are described briefly below:

Pure-line selection involves selection from the extremes for the desired phenotype. In genetic terms, a selected and inbred population is more homozygous than its wild relatives. Pure-line selection generally involves three distinct steps. First, a large number of selections are made from the genetically variable original population. Second, progeny rows are grown from the individual plant selections for observational purposes. Finally, when the breeder can no longer decide between lines solely on the basis of observation, he/she must turn to replicated traits comparing among remaining selections with established commercial varieties in relative yielding ability and other aspects of performance. This stage of evaluation lasts at least three years.

In contrast to pure-line selection, the most frequently employed plant breeding technique is hybridization. Hybridization demonstrated that crops such as maize could be inbred for six or seven generations until there was no further reduction in vigor or size. When these highly inbred plants were hybridized with other inbred varieties, very vigorous, large sized, large fruited plants were produced. This led to the origin of hybrid maize in 1919 (Chrispeels and Sadava 1977) then the most significant improvement in American agriculture at that time. The term "heterosis" was used to describe this phenomenon of hybrid vigor.

Instead of relying only on the introduction of genetic variability from the wild species gene pool, another alternative is the introduction of spontaneously mutations or these induced by either chemicals or radiation. The mutant obtained is tested and further selected to meet standards of an established cultivar, or used as a donor in a crossing program. This method is applicable for both modes of pollination. It has not been widely used in breeding programs since the great majority of mutations carry undesirable traits.

Normal plants are diploid. Plants with three or more complete sets of chromosomes are common, and are referred to as polyploids. The increase of chromosome sets can be induced by applying the chemical colchicine which disrupts spindle formation during cell division so that daughter chromosomes sets remain in the same cell. Thus, the chromosome number is doubled. Generally, the main effect of chromosome doubling is to increase size. Morphologically, polyploids tend to be larger than diploids, with thicker leaves, and they may respond differently to environmental conditions. Genetically, the plants may also differ. For example, an autotetraploid with two allelic forms of a gene at a locus (A and a) can have five genotypes in the population (AAAA, AAAa, AAaa, Aaaa, aaaa). This increases its genetic variability. Autopolyploids contain genes similar to their diploid progenitors whereas allopolyploids combine the gene content of two different species. The latter possesses higher potential capacity for variation.

2. Hybrid Plants

Our forefathers recognized very early that both animals and plants could be selected and crossed to reproduce and upgrade species. Although there was no record when the first hybrid plant was made, the first hybrid animal (mule) was produced between a horse and a donkey over 2,000 years ago. The first drastically improved crop species was probably a hybrid plant. Plant breeding was recently described as the selection of plants with desired traits after sexual exchange of genes by cross fertilization between two parents (Goodman et al 1987). This clearly indicates that hybridization is an essential technique in plant breeding and the resulting plants are hybrid plants. There are, in general, two types of hybrid plants; interspecific and intergeneric hybrids. Beyond this biological boundary hybridization cannot be accomplished due to sexual incompatibility. The successful breeding of wheat (McFadden 1930), tomato (Bohn and Tucker 1939), and soybean (Newell and Hymowitz 1982) are noted examples of interspecific hybridizations. The origin of many tobacco species was believed to arise from the interspecific hybridizations followed by the doubling of the chromosomes (Gray et al 1974). On the other hand, some modern crop species, such as rapseed and certain wheat strains originated in nature by hybridization between different species or genera (Simmond 1976).

The most notable and successful hybrid plant ever produced is the hybrid maize more than half a century ago (Borlaug 1983). Many improved elite hybrids with continually higher yields, increased disease and insect resistance and shorter and stronger stalks suitable for mechanical harvesting have since been developed.

The first maize hybrid plant was developed after recognition that inbreeding in maize leads to reduced vigor in the following generation and that vigor can be restored by crossing. In 1908 Shull reported that the hybrid plants produced by a cross between two different pure lines quadrupled the yield; from 20 to 80 bushels per acre (Chrispeels and Sadava 1977). By 1919, the first commercial hybrid maize was hybrid, as it is to this day (Chrispeels and Sadava 1977).

There are two technical steps in the production of hybrid maize plants; the production of desirable homozygous lines and the crossing of these lines. Hybrid maize plants produced are the result of double crossing. One parental plant possessing the unique property of cytoplasmic male sterility was sterility was used to avoid inbreeding. Currently, hybrid maize plants produced in the U.S. can have yields as high as 130 bushels per acre.

Although interspecific as well as intergeneric hybrid plants can be produced in certain species, there are still severe limitations in obtaining hybrid plants between many genera and even species because of sexual incompatibility. Sexual incompatibility, in most cases, is genetically controlled and regulated by the pollen-stigma recognition processes (Heslop-

Harrison 1978). This recognition is vitally important in preserving species. Otherwise, distinctions among plant families, genera and species would disappear. However, this incompatibility limits our ability to widen genetic variability within a given plant species because it precludes the natural genetic exchange between distant or unrelated species. Even within similar species, there may be self- or cross-incompatibility. To overcome this problem, techniques for hybridization which bypassed the sexual cross were developed.

3. Green Revolution

Success in using plant breeding principles and agricultural practices to improve crops reached its peak when high-yielding wheat and rice strains were cultivated in the 1960s (Borlaug 1983). The doubling or tripling of productivity of these important crops in Asia signaled a new agricultural revolution in the developing countries. This breakthrough in food production was termed the "Green Revolution" to describe the social, economic and nutritional impact of the new high-yielding wheat and rice strains (Chrispeels and Sadava 1977). Dr.Norman Borlaug was awarded the Nobel Peace Prize in 1970 for his contribution in breeding new high-yielding strains of cereals. Since that time the Green Revolution has been both praised and damned. In the 1960s, as use of the new strains spread rapidly in Asia, claims were made that many underdeveloped countries would soon be self-sufficient in cereal grains. However, these new high-yielding strains were heavily dependent on fertilizer and irrigation and required energy intensive investments. Thus, in the early 1970s, poor weather, high energy costs and global economic constraints slowed the progress of the revolution considerably. Thus the Green Revolution illustrates both potentials and limitations of conventional plant breeding technology.

III. TISSUE CULTURE, PROTOPLAST FUSION AND SOMATIC HYBRID PLANTS

1. Tissue Culture

In contrast to animal cells, plant cells are totipotent, each living cell is capable of regenerating into an entire plant identical to the one from which the living cell was obtained. The term "tissue culture" was used to describe any type in vitro culture of plant parts; callus, organ, embryo, anther, cell suspension, protoplast and many others. It was developed in the 1950s. Tissue culture techniques proved to be more valuable in commercial mass production of uniform progenies than in crop improvements. There are exceptions. One of the outstanding examples are the new tobacco cultivars produced by anther culture.

It should be mentioned that one of the properties of tissue culture is that the progenies produced by this technique are highly variable. This induced variability is termed somaclonal variation and is frequently used as a means to generate mutations. Such variability includes chromosome number changes, chromosome structure rearrangement, DNA amplification and de-amplification, rearrangement and mobilization, single gene mutation and alteration of gene expression. By using this approach, for example, herbicide resistant mutants were selected. It was estimated that in vitro tissue culture produced ten times more somaclonal variations than can be induced by chemical treatments. Thus, it serves as a reliable approach to induce mutations.

2. Protoplast Fusion

Protoplast fusion was developed after the successful culture of a large number of plant cells stripped of their walls. The resulting naked plant cells have been referred to as protoplast (Shepard et .al 1983). Protoplasts can be studied as a simple cellular entity like a micro-organism. The isolated protoplasts can be used for fusion, DNA uptake, to study cell walls and other cellular investigation. In the 1960s, Takebe and his co-workers prepared tobacco protoplasts for efficient viral infection experiments. Protoplasts were soon adapted for regeneration (Nagata and Takebe 1971). Success in regenerating complete plants from protoplasts eventually led to attempts to combine cells with different genetic background. In 1972, Carlson et al were the first group to succeed in fusing tobacco protoplasts from two genetically compatible Nicotiana species N. glauca and N. langsdorffii. Initially the fused products were termed parasexual hybrids (Carlson et al 1973). Later the term somatic hybrid was adopted (Shepard et al 1983).

3. Somatic Hybrids

Somatic hybrid plants are hybrids derived from the fusion of somatic cells. Since all fusions are possible in principle, regardless of the extent of genetic relatedness, somatic hybrid plants theoretically offer unlimited possibilities for genetic exchanges. The develop-ment of somatic hybrids was a new approach to plant breeding based on the unique property of plant cells. All plant cells can be isolated, cultured and induced to regenerate into complete plants identical to the parent from which these cells were obtained, a property known as totipotency. The dramatic success in producing somatic hybrids generated great excitement and great expectations in the 1970s. Hopes on creating new crops were high that this technology could permit unlimited generic exchange. Plant biologists speculated about the possibility of producing a somatic hybrid combining tomato, potato and tobacco could produce an efficient plant whose leaves, fruits and tubers could be harvested and economically utilized. To move towards this goal, many fusions between protoplasts of phylogenetically unrelated species were attempted. Notable examples as reviewed and compiled by Ouellette and Cheremisinoff (1985) included fusion between soybean and tobacco, soybean and barley, potato and tomato, carrot and parsley, sorghum and maize, and even plant and human cells (Lima-De-Faria et al 1977). In most cases, not surprisingly, these attempts failed. No progress could be made beyond the simple fusion stage since the initial fusion between any two protoplasts of different background is achievable but subsequent division and growth of the fused product is almost impossible.

The success in fusing two sexually compatible Nicotiana species led to the fusion between two sexually incompatible petunia species, P. parodii and P. parviflora (Power et al 1980). Interspecific protoplast fusion can be considered as a conceptual extension of interspecific sexual crosses of incompatible species, a substantial contribution. Beyond this plant breeding technology, success of intergeneric protoplast fusion was very limited. Many fertile intergeneric somatic hybrid plants were generated in the Brassicaceae family (Glimelius et al 1991). These have been used for backcrossing to the cultivated species. Nevertheless, the well publicized somatic hybrids between potato and tomato deserves special mention.

Potato and tomato are members of the same family (Solanaceae) but different genera. Potato belongs to the genus Solanum and tomato to Lycopersicon and their somatic chromosome numbers are 48 and 24 respectively (Shepard et al 1983). They are, of course, not sexually compatible. In 1978, Melchers et al were the first to produce an intergeneric somatic hybrid between potato and tomato. They fused protoplasts between cultured diploid potato line and tomato leaf cells. Hybrid plants displayed morphological features of both parents like most hybrids. The hybrid plants were named "pomatoes" (Melchers et al 1978). Some hybrid plants formed "tuber-like stolons"; however, none set fertile flowers or fruits or produced true tubers (Shepard et al 1983). Somatic hybrid plants produced to date, such as the pomato are not of immediate economic value but they serve the starting point of a genetic introgression scheme. However, it should be mentioned that the contributions made from these fusion studies on the interaction between the nucleus and organelles (Kung et al 1975) and on the regeneration and development of cell wall and others are invaluable.

The advance from the era of hybrid plants to the era of somatic hybrid plants permitted expansion of the scope of plant breeding to both the organismic and cellular levels. Further advances in plant breeding-to the molecular -are the subject of the next section. It demonstrates that we are now entering the new era of biotechnology or genetic engineering.

IV. BIOTECHNOLOGY, TRANSGENIC PLANTS AND THE GENE REVOLUTION

1. Biotechnology

Biotechnology offers new ideas and techniques applicable to agriculture. It uses of the conceptual framework and the technical approaches of molecular biology to develop commercial processes and products. With the rapid development of biotechnology, agriculture has moved from a resource-based to a science-based industry. Indeed, plant breeding has been dramatically broadened by the introduction of genetic engineering techniques based on knowledge about gene structure and function. This has ushered agriculture into a new era where it joins ranks with the most sophisticated of biological sciences in using molecular approaches whose promise to yield new generation of plants of any desirable trait is matched by their widely demonstrated potential in medicine. The technology required for engineering transgenic plants is considerably more sophisticated than the one for producing somatic hybrid plants by cross fertilization. In 1983, direct gene transfers became possible using recombinant DNA technologies (Fraley 1989). The newly acquired ability to transfer genes among organisms without sexual crossing provides breeders with new opportunities to improve the efficiency of production and to increase the utility of agricultural crops. Plants with new traits such as herbicide resistance have been genetically engineered using genes from unrelated organisms. Scientists are also trying to improve the quality of our agricultural products by altering the nutritional value of proteins and oils. However, it should be emphasized that the biotechnology is not a substitute or replacement for the conventional breeding methods. Rather, it can improve on past, conventional methods. The major differences between conventional breeding and biotechnology lie neither in goals nor processes, but rather in speed, precision, reliability and scope.

By using recombinant DNA technology many transgenic organisms have been engineered since 1985. Table 1 lists over 40 transgenic plants produced to date. The list is growing rapidly because the technology for inserting and expressing genes in plants is in hand, and more genes are being identified, together with the traits they control.

The majority of plant biotechnology research thus far has focused on developing plants that resist insects and diseases, frost damage and tolerate certain herbicides. Insect resistance research has centered on inserting a gene obtained from the bacteria Bacillus thuringiensis (Bt) into plants or bacteria. The gene produces a toxin that is fatal to insects.

Efforts have concentrated on the use of Bt because it has been used commercially in agriculture for 25 years and dissipates rapidly in the environment.

Herbicide resistance research has focused on developing tomato and tobacco cultivars that are resistant to herbicides (Fraley 1989) such as glyphosate (Roundup, Monsanto, St. Louis, MO). Environmental groups have expressed concern about the benefits from this research since it is not clear whether it will lead to increased or decreased use of herbicides. Commercialization of these plant technologies may be a few years away as a result of economic incentives, environmental regulation, and public acceptance. For example, a major economic consideration will be whether or not cultivars produce the fruits that meet the standards of processors.

Plant technologies that involve multigenic traits such as nitrogen fixation and stress tolerance are even further away. Researchers are concerned because plants that are engineered to fix nitrogen may use so much energy in doing it that yields may fall dramatically. Much more needs to be learned about genetic structures, gene function, and gene regulation before these applications are achievable and marketable

2. Transgenic Plants

There are four essential components, in addition to desirable genes, for a successful production of transgenic plants. They are suitable vectors, selectable markers, efficient techniques for transformation and regeneration. It is quite obvious that the dramatic progress made recently in the development of gene transfer system (Gasser and Fraley 1989) makes it possible to produce the numerous transgenic plants (dicot and monocot). our current ability to insert foreign genes into plant cells and tissues opportunities to plant breeders unmatched in history. This allows for explosive expansion of our understanding in the field of plants in ways unthinkable only a decade ago. This remarkable progress is clearly reflected by the steady increase in numbers of transgenic plant species produced (Roger 1988). Their number increased to 30 in 1989 (Gasser and Fraley 1989, Ratner 1989) and over 40 a year later (Table1).

A brief review of the literature on transgenic plants indicates that transgenic plants have been engineered with several purposes; developing new techniques for transformation, basic study on a specific gene and crop improvement for a specific cause. Among the 50 publications on transgenic plants reviewed by the author most focus still on the development of suitable methods for the transformation and regeneration of major crops. The plants transformed were predominantly dicot particularly from the family Solanaceae with only a few monocots and woody plants

The first transgenic plants were engineered in the 1980s. Hall, Kemp and their coworkers transferred the β-phaseolin gene from bean to sunflower and tobacco plants (Murai et al 1983). Independently, several transgenic tobacco plants were produced to express foreign genes engineered by the Agrobacterium tumefaciens vectors (Horsch et al 1984, De Block et al 1984). Now, a variety of free DNA delivery systems are available. This was followed by the highly publicized transformation of luciferase gene from firefly to tobacco (Ow et al 1986). The early transformation experiments often utilized plant protoplasts as the recipient cells; the subsequent development of transformation methods based on regenerable explant

(Horsch et al.1985) such as leaves, stems and roots simplified significantly the transformation techniques that are widely used today. The recent breakthrough in the regeneration of transformed monocot plants rice and maize, has removed a major obstacle in improving cereal crops. Production of transgenic plants in most laboratories is now a routine operation. Many papers on transgenic plants are published each year.

3. Gene Revolution

The current advancement of biotechnology has ushered agricultural research into a new era of transgenic organisms or gene revolution. It is notable that as the range of technologies available for crop improvement expands, the role of technological progress is rapidly increasing. It requires infinite knowledge of plant biochemistry and physiology, and development at the molecular level. This is a striking departure from the traditional view of agricultural research. Thus, we have advanced from the hybrid plants to the transgenic plants, or from the green revolution to the gene revolution.

V. FUTURE PROSPECTS

Crop improvement represents a continuous efforts to seek new ways and to solve problems associated with conventional methods of improving crops. The application of genetic principles in breeding programs has been very successful and will continue to be. It formulated the basis for the "Green Revolution." This was further complemented by the development of tissue culture and protoplast fusion techniques. The combination of conventional breeding approaches and tissue culture techniques has opened many avenues for crop improvements which were not available previously. Currently, genes used for plant transformation are not of plant origin. It is anticipated that this may change in a few years. Major efforts are underway to identify the genes of disease or pest resistance. Our next giant leap in crop improvement will be the transfer of plant genes into plants for fortifying their natural defense capability. The ultimate goal is to construct a stable agroecosystem through biotechnology that supports high productivity without constant environmental or chemical interventions.

ACKNOWLEDGEMENT

Greg Silsbee's help in preparing this manuscript and Dr.Shyam Dube's review are greatly appreciated.

TABLE I. TRANSGENIC PLANTS PRODUCED
Monocotyledonous Plants

Asparagus	Orchard Grass
Maize	Rice
Millet	Rye
Wheat	

Dicotyledonous Plants

Legumes	Special Crop
Alfalfa	Cotton
Clover	Flax
Peas	Lotus
Soybean	Sugar Beet
Moth bean	Sunflower

Solanaceae	Woody Plants
Eggplant	
Petunia	Apple
Potato	Pear
Tobacco	Poplar
Tomato	Walnut

Vegetable and Fruit Crops	Special Plant
Cabbage	Arabidopsis
Carrot	
Cauliflower	Other Groups
Celery	Medicago varia
Cucumber	Vigna aconitifolia
Horseradish	Stylesanithes humilis
Lettuce	Morning Glory
Rape	Kalanchoe
Grape	
Muskmelon	
Strawberry	

Sources: 1 Gasser and Fraley 1989   2. Fraley 1989
3. Ratner 1989   4. Roger 1988

REFERENCES

Bohn, G.W. and Tucker, C.M. (1939). Science 89, 603-605.

Borlaug, N.E. (1989) Science 219, 689-693.

Carlson, P.S. (1973) Proc. Natl. Acad. Sci. USA, 70, 598-602.

Carlson, P.S. Smith, H.H. and Dearing, R.D. (1972). Proc. Natl. Acad.Sci, USA, 69, 2292-2294.

Chrispeels, M.J. and Sadava, D. (1977) Plants, Food and People. W.H. Freeman and Co., San Francisco.p.192.

De Block, M. Herrera-Estrella, L. Van Montagu, M. Schell, J. and Zambryski, P. (1984). EMBO J. 31, 681-686.

Fraley, R. 91989) Plant Biotechnology, Butterworths, Boston. Ed.S.D. Kung and C.J. Arntzen, pp. 395-407.

Gasser, C.S. and Fraley, R.T. (1989) Science 244, 1293-1299.

Glimelius, K., Fahlesson, J. Landgren, M.Sjodin, C. and Sundberg, E. (1991). Trend in Biotech. 9, 24-30.

Goodman, R.M. Hauptli, H. Crossway, A. and Knaup, V.C. (1987). Science 236, 48-54.

Gray, J.C.Kung, S.D. and Wildman, S.G. (1974) Nature 525, 226-227.

Heslop-Harrison, J. (1978) Cellular Recognition Systems in Plants. Studies in Biology, Edward Arnold. London.

Horsch, R.B. Fraley, R.T., Rogers, S.G. et al (1984). Science 223, 496-498.

Horsch, R.B. Frey J.E. Hoffmann et al (1985)Science, 227, 1229-1231.

Kung, S.D., Gray, J.C., Wildman, S.G., and Carlson, P.S. (1975). Science 187, 353-355.

Lima-De-Faria,A.Eriksson, T. and Kjellen, L. (1977). Hereditas 87, 57-61.

McFadden, E.S. (1930) J.Am. Soc. Agron. 22, 1050-1054.

Melchers, G., Sacristan, M.D. and Halder, A.A. (1978) Carlsberg Res. Commen. 43, 203-

Murai, N. Sutton, D.W. Murray, M.E. et sl (1983) Science 22, 476-482.

Nagata, T.and Takebe, I. (1971) Planta. 99, 12-16.

Newell, C.A. and Hymowitz, R. (1982). Crop Sci 22, 1061-1066.

Ouellette, R.R. and Cheremisinoff, P.N. (1985) Application of Biotechnology. Technormic pub. Co. Lancaster, PA.

Ow, D.W. Wood, K.U. DeLuca, M et al. (1986) Science 234, 856-859.

Power, J.B. Berry, S.F., Champman, J.V. and Cocking, E.C. (1980). Theor. Appl. Genet. 57, 1-6.

Ratner, M. (1989) Biotechnology. 7, 337-341.

Rogers, S. (1988) Proceedings of Transgenic Plant Conference. Annapolis, MD, pp. 31-40.

Shepard, J.F. Bidney, D.F. Basby, T. and Kemble, R. (1983) Science 219, 683-688.

Simmonds, N.W. (1976) Evaluation of Crop Plants, ed. Longman, N.Y.

THE ROLE OF LECTIN IN ASSOCIATION BETWEEN RICE AND NITROGEN - FIXING BACTERIA

S. Chaopongpang, S. Pornpattkul, C. Pitaksutheepong,
J. Limpananont*, P. Chaisiri and J. Boonjawat

Department of Biochemistry , Faculty of Science,
*Faculty of Pharmaceutical Science
Chulalongkorn University, Bangkok 10330 Thailand

ABSTRACT

Lectins are multivalent carbohydrate-binding proteins and glycoproteins. In rice and other cereal crops, lectins are well characterized for their binding specificity to N-acetyl-glucosamine or its oligomers. Although the natural function of lectin remains unknown, several possible functions of cereal lectins have been proposed including an involvement in plant-microbial interaction. We found that rice lectins are not exclusively confined to embryo, but also in seedling leaf and root, especially at the root tip, and opening stoma of young leaf, which are preferent adhesion sites of microbes. Lectin contents varied with developmental stage and environmental factors. Different rice varieties contain different amount of lectin and respond differently to inoculation by associative nitrogen-fixing *Klebsiella* R15 and R17. Better plant growth promotion and associative nitrogen-fixation were observed in high lectin variety. Colonization of bacteria on the root surface increased the root lectin. Our results support for the potential application of nitrogen-fixing *Klebsiella* for promotion of growth and associative nitrogen fixation in rice, *Oryza sativa* L.

INTRODUCTION

Many purified plant seed agglutinins or lectins which are proteins and glycoproteins that bind carbohydrate of different sugar specificities have been reported as determinant of host-plant binding to its specific symbiont (Bohlool and Schmidt,1974; Dazzo and Hubbell,1975). Trifoliin A, a lectin found in the root exudate and the root surface of white clover has been reported to bind specifically to *Rhizobium trifolii* (Truchet et al. 1986). By introducing the pea lectin gene into white clover roots using *Agrobacterium rhizogenes* as vector, the

Biotechnology and Environmental Science: Molecular Approaches
Edited by S. Mongkolsuk et al., Plenum Press, New York, 1992

81

clover roots of transgenic plant become hairy and can be nodulated by a *Rhizobium* strain usually specific for pea (Diaz et al.,1989). In rice cv. RD7, lectins purified from seedling root, embryo and bran were demonstrated to be involved in the adhesion between rice root and *Klebsiella* strain R15 and R17 (Boonjawat et al. 1988).

In this report, we will summarize the results of our studies in the following aspects : 1) rice lectin content during seedling development, 2) variation of lectin content in different varieties of local rice cultivars, 3) the effect of ammonium ion on lectin content, 4) the relationship among colonization potential, nitrogen-fixing potential, plant vigor index and lectin content, and 5) the effect of bacterial inoculation on rice lectin content in high lectin cultivar, and low lectin cultivar.

MATERIALS AND METHODS

Breeder seeds of rice (*Oryza sativa* L.) subspecies indica were obtained from the Rice Germplasm Bank, Pathumthani Rice Research Center, Department of Agriculture, or from the Department of Rice Research, Ministry of Agriculture and Cooperative. Rice seeds were surface sterilized and grown in sterile water under aseptic condition.

Klebsiella strain R15 and R17, the nitrogen-fixing rhizobacteria were maintained in rich medium (RM, Luria et al.,1960) containing 15% v/v glycerol at -70°C, and subcultured on RM agar and RM broth as starter. A 1-3% inoculum from starter was inoculated in nitrogen-free (NF) medium (Dobereiner, 1977) incubated at 37°C in a rotary shaking water bath. Cells were harvested and washed with normal saline or phosphate buffer saline (PBS) before use as inoculant at specified concentration.

Development of lectin antibody and ELISA procedure to determine lectin content in rice seedlings were modified from Raikhel and Pratt, 1987 and Christensen et al. 1986, as shown in Figure 1. Protein was determined according to Bradford (1976). Development of indirect ELISA procedures to assay associative bacteria on rice roots were modified from Fuhrmann and Wollum II, 1985 as summarized in Figure 2.

Nitrogen-fixing potential of rice plant in association with bacteria was assayed by acetylene reduction activity (ARA) using 3 rice seedlings grown in 30 ml modified Weaver medium (Weaver et al. 1975).

Plant vigor index (Purushothaman, 1987) is the length of shoots (cm) plus roots length (cm) and multiplied by per cent germination.

Abbreviations - ARA, acetylene reduction activity; COM-IND-ELISA, competitive indirect enzyme-linked immunosorbent assay; NF, nitrogen-free; PBS, phosphate buffered saline; PTN, 10 mM phosphate buffer pH 7.4 containing 0.15 M NaCl, 0.05% Tween-20, and 0.02%NaN$_3$; Rice cultivars, cv. NMS4 (Nangmol S4), KTH 17 (Khaotahaeng 17), SPBR60 (Supanburi 60), BMT470 (Basmati470), KDML105 (Khaodokmali 105), RD# (Rice Dept.#).

Plate coating by 0.2% ovalbumin (lectin receptor) 100 μl/well

(in 0.1M Na_2CO_3 pH9.6)

↓ incubate 4°C, overnight

Wash with PTN* buffer

Add 100 μl of tissue extract at a proper dilution

↓ incubate 37°C, 30 min

Wash with PTN

Add 100 μl of the first antibody (antilectin)

↓ incubate 37°C, 45 min

Wash with PTN

Add 100 μl of enzyme-linked antirabbit-immunoglobulin

↓ incubate 37°C, 45 min

Wash with PTN

Add 100 μl of 1 mg/ml p-nitrophenyl phosphate (substrate)

↓ incubate 37°C,1h

Stop reaction by adding 50 μl 1 N NaOH

Read OD_{405} with an immuno-assay reader

*PTN : 10 mM phosphate buffer, pH 7.4, containing 0.15 M NaCl, 0.05% Tween-20, and 0.02% NaN_3

Figure 1. Protocol of indirect ELISA for lectin determination

Plate coating by cells antigen 100 μl/well

↓ incubate 5°C, overnight

Wash with PTN buffer

Add 100 μl of soluble test antigen mixed with antiserum

↓ incubate 30°C, 1h

Wash with PTN

Add 100 μl of enzyme linked antirabbit immunoglobulin

↓ incubate 37°C, 3h

Wash with PTN

Add 100 μl of 1 mg/ml p-nitrophenyl phosphate (substrate)

↓ incubate at room temperature 50 min

Read OD_{405} with an immuno-assay reader

Figure 2. Protocol of COM-IND-ELISA for *Klebsiella* spp.

RESULTS AND DISCUSSION

Lectin contents during development of rice seedlings

Lectin levels in leaf and root homogenates of rice cv. RD7 grown under dark and light conditions were determined during 4-7 days after germination. Table 1 shows that lectin level (ng/mg total protein) and total lectin content (ng/100 plants) are highest on day 4 after germination in the dark. The leaf tissues contain higher amount of lectin comparing to roots, and the level of leaf lectin reduces significantly (2.5 fold) in lighted condition, where root lectin remains more or less similar. By using immunofluorescent labelling, we found that leaf lectin was localized on hydathode and opening stoma, and root lectin accumulated at the root tips and roothair tips (data not shown). The presence of leaf lectin suggests for the possibility of spraying the nitrogen-fixing bacteria and establish phyllospheric attachment on the leaf surface of rice seedlings as well as attachment on the root surface.

Table 1. Effect of light and developmental stage on lectin content

Rice seedlings cv. RD7 were grown in sterile water, then extracted with PBS, and assayed for lectin content by ELISA procedure shown in Fig. 1.

Time (day)	Lectin(ng/mg Protein)				Total lectin (ng/100 plants)			
	Light		Dark		Light		Dark	
	Leaf	Root	Leaf	Root	Leaf	Root	Leaf	Root
4	109±10*	17±1	210±41*	15±2	893±80*	44±1*	2628±552*	55±6*
5	40±1	14±1	127±12	19±1	461±9	32±2	1367±132	38±2
6	39±4	16±1	51±4	11±1	501±53	39±2	603±49	15±1
7	9±2	12±1	17±2	12±2	175±44	23±2	153±20	28±5

Values presented are mean of 6 measurements.

*Significant different between light and dark condition, by t-test at 95% confidence.

Variation of lectin content in different rice cultivars

By using 4 day old rice seedlings germinated in the dark, the lectin level in root extracts (with 1 ml PBS) of 8 cultivars of rice are shown (Table 2) to vary from 6-31 ng/mg total protein, and 5-92 ng/100 plants. It is noted that among these 8 cultivars tested, NMS4 and RD7 may be classified as rather high-lectin cultivars, and KDML as a low-lectin cultivar. The total lectin content in leaf decreased in the same order as in the root, but lectin concentration per total protein shows different order because each cultivar may contain different percentage of total protein.

Table 2. Lectin content in leaf and root of 8 rice cultivars

All cultivars were grown under dark condition for 4 days, root and leaf were extracted with PBS, and assayed for lectin content by ELISA procedure shown in Fig. 1.

Rice cultivar	Lectin/Total protein (ng/mg P)		Total lectin content (ng/100 plants)	
	Leaf	Root	Leaf	Root
NMS4	228±61	31±7	4,230±1,135	92±20
RD7	210±42	15±2	2,628±522	58±6
RD23	307±29	29±2	2,405±224	34±2
SPBR60	151±19	13±1	1,776±228	33±3
KTH17	32±2	8±1	859±150	31±2
BMT470	136±14	19±1	773±80	24±1
RD25	115±17	10±1	771±104	17±1
KDML105	59±15	6±0	598±151	5±0
Mean ± S.D.	155±85	16±9	1,758±1,167	36±25

Values presented are mean of 6 measurements ± S.D.

Effect of exogenous NH4 ion on lectin content

Excess NH_4Cl (20 mM) resulted in about 2-fold reduction in lectin concentration and content in the root and leaf of 4-day old rice seedlings cv. RD7 grown in the dark (Table 3). This result suggests that lectin biosynthesis might be repressed in the presence of excess exogenous combined nitrogen.

Table 3. Effect of exogenous ammonium ion on lectin content

Rice seedlings cv. RD7 were germinated in sterile water or supplemented with NH_4Cl at 2 and 20 mM for 4 days under dark condition before lectin determination by ELISA.

[NH₄Cl] (mM)	Lectin/Total protein (ng/mg P)		Total lectin content (ng/100 plants)	
	Leaf	Root	Leaf	Root
0	228±61	31±7	2,628±552	92±20
2	212±5	24±1	1,681±57*	102±4
20	173±1*	13±1*	1,623±16*	49±2*

Values presented are mean of 6 measurements.

*Significant level of lectin content between 0 and 20 mM NH_4Cl.

Statistical analysis by t-test at 95% confidence.

The relationship among colonization potential, nitrogen-fixing potential, plant vigor index and lectin content

Different cultivars of rice seedlings were grown aseptically in NF-Weaver medium made semi-solid with 0.5% Noble agar, and inoculated with NF-grown *Klebsiella* R15 and R17

(5x 10^8 cells/tube) on day 2 after germination. Normal saline was added instead of bacteria for control tubes. After inoculation, ARA, root colonization and plant vigor index were measured in 12 replicative tubes, each contained 3 seedlings in 30 ml semisolid medium, with or without bacteria. Under these conditions, the maximum colonization potential, nitrogen-fixing potential and plant vigor index over control tubes were observed on day 12 after inoculation. The relationship among these 3 criteria for compatible association between N_2-fixing bacteria and rice were therefore related with known lectin levels as shown in Table 4.5 and 6. Among 4 rice cultivars, in which RD7 contains the highest amount of lectin per plants, this cultivar seemed to support bacterial colonization better than KDML, the low-lectin cultivar (Table 4), and the net gain in associative nitrogen fixation can be observed in only 2 rice cultivars, RD7 and RD25 under this experimental conditions as shown in Table 5. The plant vigor index was higher in the associative pairs showing both N_2-fixing activity, and colonization greater than 10^9 cells/g root dry weight, and the least benefit of *Klebsiella* R15 and R17 inoculation can be noted in KDML, the low-lectin rice cultivar (Table 6).

Table 4. The relationship between colonization potential and root lectin content.

Rice seedlings were grown in modified Weaver medium, 3 seedlings per tube, and inoculated with *Klebsiella* R15 or R17, 5x10^8 cells/tube on day 2, and normal saline was added in control tubes. The colonization potential was the number of bacterial cells associated with the rice roots determined by competitive indirect ELISA (COM-IND-ELISA) procedure shown in Fig. 2.

Rice cultivar	Root lectin* (ng/mg P)	Root lectin* (ng/100 plants)	Log colonization potential (log cells/g dry wt. root) R15	Log colonization potential (log cells/g dry wt. root) R17
RD7	15	58	9.32	9.32
RD23	29	34	8.23	8.54
RD25	10	17	9.16	9.17
KDML105	6	5	8.49	8.48

*Values of root lectin were from Table 2.

Table 5. The relationship between nitrogen fixation potential and root lectin content.

Rice seedlings were grown in modified Weaver medium, 3 seedlings per tube, and inoculated with *Klebsiella* R15 or R17, 5x10^8 cells/tube on day 2, and normal saline was added in control tubes. Nitrogen fixation potential was the net gain in ethylene (μmol/g root dry weight) produced in the head space over control tubes (12 replicate tubes for each set) measured on day 12 after inoculation, which was the maximal ARA activity.

Rice cultivar	Root lectin* (ng/mg P) R15	Root lectin* (ng/100plants) R17	N_2-fixing potential (μ mol C_2H_2/g dry wt. root)	N_2-fixing potential (μ mol C_2H_2/g dry wt. root)
RD7	15	58	111.0	82.8
RD23	29	34	-	-
RD25	10	17	5.0	0.3
KDML105	6	5	-	-

*Values of root lectin were from Table 2.

Table 6. The relationship between plant vigor index and root lectin

Rice seedlings were grown in modified Weaver medium, 3 seedlings per tube, and inoculated with *Klebsiella* R15 or R17, 5x10⁸ cells/tube on day 2, and normal saline was added in control tubes. Plant vigor index was determined on day 12 after inoculation, which was the maximal growth under this experimental condition.

Rice cultivar	Root lectin* (ng/mg P)	(ng/100plants)	Plant vigor index (cm.%) R15	R17
RD7	15	58	2,588	2,210
RD23	29	34	920	918
RD25	10	17	1,555	1,575
KDML105	6	5	253	793

*Values of root lectin were from Table 2.

The effect of bacterial inoculation on lectin content

When inoculum of *Klebsiella* R15 (10⁹ cells/plant) were added in 7-day old seedlings of rice cv. NMS4, and KDML and assayed for lectin content on day 14, Table 7 shows that the root lectin concentration and total lectin content increased in both inoculated experiments. This result suggests that bacterial inoculation might induce more lectin synthesis in the associative rice plant.

Table 7. Effect of bacterial inoculation on lectin content

Rice seedlings were grown in sterile water, bacteria R15 were added on day 7 after germination (10⁹ cells/plant, and normal saline was added to control plants. On day 14, root and leaf from 100 plants were extracted with PBS and assayed for lectin content by ELISA.

Rice cultivar	Treatment	Root lectin content (ng/mg protein)	(ng/100plants)
High lectin NMS4	Control	57±5	166±11
	+R15	76±17	299±51
Low lectin KDML105	Control	21±1	21±1
	+R15	28±2	44±3

Values presented are mean of 6 measurements ± S.D.

In conclusion, the lectin content in rice can be affected by endogenous factors such as type of tissue, age, variety, and exogenous environmental factors such as light/dark, combined nitrogen and bacterial colonization. These variable parameters should encounter for inconsistent outcome of rice-*Klebsiella* association. Our results still support for the hypothesis that lectin of some wetland rice cultivars should play important role in enhancing adhesion and colonization of associative *Klebsiella* spp., but should not be involved directly in enhancing the associative nitrogen fixation. However the colonization of these rhizobacteria could by some means induce lectin synthesis in the root. An increase in plant vigor index in the absence of nitrogen fixation due to rhizospheric colonization indicated that, these rhizobacteria might contribute some plant growth promoting substances. In case of *Azospirillum*, an ubiquitous Genus of associative N_2-fixing rhizobacteria, it has been demonstrated that promotion of root growth in many cereal crops, including forage grass, legumes, and tomatoes (Dobereiner and Pedrosa, 1987) can contribute to increased yields and exerted beneficial effects at the field trials level. Identification of indole acetic acid and related substances of auxin have been reported in *A. brasilense* (Fallik et al., 1988). We have observed that inoculation by *Klebsiella* R15, and R17 increased the total number and total length of adventitious roots and roothairs of rice seedlings as reported in *Sorghum bicolor* inoculated by *A. brasilense* (Okon et al., 1989). It is hoped that by optimization of the inoculum size, time of inoculation, and cultivars of rice using lectin and plant vigor index as criteria, should assist in successful program of enhancing associative nitrogen fixation in some varieties of indica rice.

ACKNOWLEDGEMENTS

This work was supported by the National Center of Genetic Engineering and Biotechnology (NCGEB 32-2#7), and the Rockefeller Foundation (RF 89026 #22).

REFERENCES

Bohlool, B.B. and Schmidt, E.L. 1974. Science 185,269-271.

Boonjawat, J., Limpananont, J., Horisberger, M. 1988. Nitrogen Fixation: Hundred Years After. Proceedings of the 7th International Congress on Nitrogen Fixation, p. 792. Eds. H. Bothe, F.J. de Bruijn, and W.E. Newton. Gustav. Fischer. Stuttgard: New York Publishers.

Bradford, M.M. 1976. Anal. Biochem. 72,248-254.

Christensen, T.M.I.V., Diaz, C.L. and Kijne, W.J. 1986. Lectins.

Biology, Biochemistry, Clinical Biochemistry, 5,31-38. Ed. T.C. Bog-hansen, G.A. Spengler, Berlin/New York: De Gruyter.

Dazzo, F.B. and Hubbel, D.H. 1975. Appl. Microbiol. 30,1017-1033.

Diaz, C.L., Melcher, L.S., Hooykass, P.J.J., Lugtenberg, B.J.J. and Kijne, J.W. 1989. Nature 338, 579-581.

Dobereiner, J. 1977. Advances in Agronomy 29,11.

Dobereiner, J. and Pedrosa, F.O. 1987. Nitrogen-fixing Bacteria in Non-leguminous Crop Plants, p. 168, Springer, Berlin.

Fallik, E., Okon, Y., Epstein, E., Goldman, A. and Fischer, M. 1988. Soil Biol. Biochem. 21,147-153.

Fuhrmann, J. and Wollum II A.G. 1985. Appl. Environ. Microbiol. 49,1010-1013.

Luria, S.E., Adams, J.N. and Teng, R,C. 1960. Virology 12,348-390.

Okon, Y., Sarig, S. and Blum, A. 1989. Recent Advances in Microbial Ecology, Proceedings of the 5th International Symposium on Microbial Ecology, p.196-200. Eds. T. Hattori, Y Ishida, Y. Maruyama, R.Y. Morita, A. Uchida, Japan Scientific Societies Press.

Purushothaman, D., Oblisami, G. and Balasun, C.S. 1976. Madras Agric. J.63,595-599.

Raikhel, N.V. and Pratt, L.H. 1987. Plant Cell Reports 6,141-149.

Truchet, G.L., Sherwood, J.E., Pankratz, H.S. and Dazzo, F.B. 1986. Physiol. Plant 66,575-582.

Weaver, P.K., Wall, J.D. and Gest, H. 1975. Arch. Microbiol. 105,207-216.

ANALYZING PATHOGEN POPULATIONS USING MOLECULAR MARKERS

Rebecca J. Nelson

International Rice Research Institute, P.O. Box 933, 1099 Manila, Philippines

INTRODUCTION

High-yielding, semi-dwarf cultivars of rice (*Oryza sativa* L.) are widely planted in many developing countries. Intensified production, involving increased fertilizer use and more rice crops per year, has led to higher yields, but has also been associated with increased susceptibility to diseases and insects. Susceptibility of crops to pests may lead to chronic or unpredictable losses in yield. Pesticides have been widely used to stabilize yields, but such chemicals may be unavailable or prohibitively expensive and may be hazardous to the environment. As an alternative to chemical control, host plant resistance and disease management practices can provide crop protection with minimal use of pesticides.

Molecular techniques provide powerful tools that can contribute to environmentally sound approaches to crop protection. Disease results from the interaction between host plant and pathogen under favorable environmental conditions. The methods of molecular biology have permitted progress towards understanding and manipulating both partners in the plant-pathogen relationship.

On the host side, resistance can be conferred through genetic engineering. Marker-assisted breeding may make it possible to manipulate resistance genes and to more efficiently incorporate quantitatively-inherited resistance. On the pathogen side, techniques for DNA analysis can be used to analyze the structure and dynamics of pathogen populations. This paper focuses on the application of molecular markers to population studies for two important pathogens of rice, and on the ways in which such information can be useful for crop protection.

PATHOGEN POPULATION BIOLOGY

For many diseases, plant breeders are able to incorporate resistance genes into crop plants. Unfortunately, pathogen populations often adapt quickly to the resistant cultivars. Specialized pathogenic strains are grouped into races or pathotypes, defined by the rice cultivars they can infect. In the case of the rice blast fungus, hundreds of physiologic races have been described (Ou, 1980), and cultivars with single resistance genes are often overcome in one or a few years (Lee and Cho, 1990).

Biotechnology and Environmental Science: Molecular Approaches
Edited by S. Mongkolsuk et al., Plenum Press, New York, 1992

89

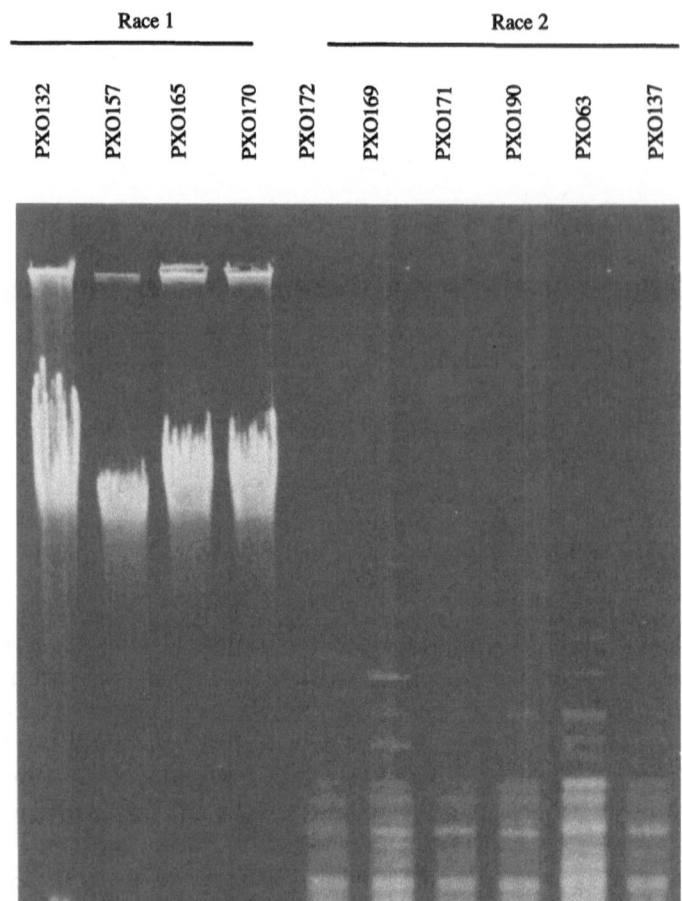

Fig.1 Agarose gel stained with ethidium bromide, showing genomic DNAs of isolates from races 1 and 2 of *Xanthomonas campestris* pv. *oryzae* after incubation with the restriction enzyme *PsfI*.

Pathogens evolve to overcome resistance because their populations are variable and respond to selection pressure. The variability of a pathogen population is determined by factors such as its mutation rate and the extent of migration (McDonald et al., 1989). The selection pressure is determined by the population structure of the host and by the environment; selection is most intense when uniform genotypes are planted over large areas under disease-prone conditions. In addition, some forms of resistance may be more durable than others (Parlevleit, 1983).

In order to breed for stable forms of resistance and to develop disease management strategies that minimize the selection pressure on pathogen populations, it is important to understand the diversity and evolutionary capacities of pathogen populations. Restriction fragment length polymorphisms (RFLPs) can act as molecular markers to differentiate among strains and among evolutionary lineages. In the work described below, we have studied aspects of pathogen population biology using repetitive DNA probes that give characteristic "fingerprints" for different strains of two important rice pathogens. These molecular markers are useful for allowing the selection of diverse pathogen strains for use in screening during plant breeding. RFLPs also provide markers for use in analyses of pathogen population dynamics.

BACTERIAL BLIGHT PATHOGEN

Bacterial blight, the most important bacterial disease of rice, is caused by *Xanthomonas oryzae* pv. *oryzae* (Xoo). The causal agent was previously classified as *Xanthomonas campestris* pv. *oryzae* (Ishiyama 1922) Dye (Ou, 1985), and has been recently reclassified (Swings et al., 1990). Host resistance is the most effective and economical approach to the control of bacterial blight. Resistant cultivars, however, may be overcome by virulent races of the bacterium. This may happen very rapidly (e.g. Ezuka and Sakaguchi, 1978). More gradual loss of resistance may be due to shifts in the bacterial population structure in response to planting of a resistant cultivar. When cultivars containing a single major gene for resistance to the disease were widely planted in the Philippines, the frequency of a pathogenic race adapted to cultivars carrying the resistance gene increased dramatically (Mew and Vera Cruz, 1985), from 10% to 90% over a ten-year period.

Physiological specialization of the pathogen has been demonstrated in Japan (Ezuka and Horino, 1974). Yamamoto and Ogawa (1988) defined 26 distinct races of the pathogen among a large number of isolates from seven South East Asian countries. Six pathogenic races have been identified among Philippine isolates, using five rice cultivars as differential hosts (Mew, 1987).

The genome of Xoo contains multiple repetitive DNA elements (Leach et al, 1989). To evaluate the population structure of Xoo in the Philippines one such repetitive DNA element, designated pJEL101, was used to probe restriction digests of a collection of Philippine strains of the pathogen (Leach et al., manuscript in preparation). Considerable variation was observed both within and among the race groupings. The RFLP types were associated strongly with race: none of the RFLP types were found in more than one race. Numerical analysis of the RFLP data suggested certain evolutionary affinities among the strains within and between races.

The objectives of our work on Xoo RFLPs have been

a. To understand the evolution, diversity, and geographical distribution of variation of Xoo in the Philippines;

c. To determine whether independently-derived RFLP data (e.g., data deriving from different probes or analytical methods) result in similar conclusions regarding evolution and diversity of a single population;

d. To develop a relatively rapid and inexpensive method for RFLP analysis of Xoo populations, so that the analysis could be done on larger sample sizes and in resource-limited conditions such as those often encountered in developing countries; and

e. To assess the diversity of host and pathogen populations in a traditional agroeco-system, and later to compare this with the bio-diversity of the "modern" agroecosystem.

To examine the evolution and diversity of the pathogen, we have isolated a set of putative transposable elements from the genome of Xoo. Using the pL3SAC transposon trapping vector (Kearney and Staskawicz, 1990), five inserts distinguishable by restriction digestion were obtained. Two of these showed homology to the pJEL101 element described by Leach et al. (Leach et al., 1989, and Leach et al., manuscript in preparation). One of the other elements, designated pTnX$_1$, was used to analyze a collection of Xoo isolates that includes those previously analyzed by Leach et al. using pJEL101 (manuscript in preparation).

Table 1. Numbers of RFLP types defined by three analyses for a set of 82 isolates of *Xanthomonas oryzae* pv. *oryzae* from the Philippines. A collection of strains was analyzed using two repetitive probes (both putative transposable elements) and using *Pst*I analysis.

Race	No. Isolates	No. RFLP types		
		pJEL101	pTnx-1	*Pst*I
1	19	7	6	1
2	18	5	2	12
3	13	7	7	4
4	4	1	1	1
5	20	3	7	9
6	8	2	1	2
Total	**82**	**25**	**17**	**29**

Total number of RFLP types differentiated by pJEL101, pTNX, & *Pst* I = 54

In comparison with the results obtained by Leach et al. (manuscript in preparation) using pJEL101, a lower overall level of diversity was obtained using pTnX$_1$ RFLP types were shared among two or three races. Race 3 shared pTnX$_1$ RFLP types with races 1, 2, 4 and 5.This supports the conclusions of Leach et al. (in preparation), who found that race 3 appears to be closely related to the other races of the pathogen. It is not clear whether race 3 is a source of variation, or whether it is the "default" race to which other races frequently mutate.

RFLP analysis of Xoo populations can provide information about population structure and race evolution. Southern blot analysis is, however, expensive and time-consuming, and may not be feasible in many national rice research programs. Fortunately, polymorphisms between strains can be detected when Xoo DNA samples were digested with the restriction enzyme *Pst*I, run on agarose gels and stained with ethidium bromide (Raymundo et al., 1990)

While no pJEL101 RFLP type was found in more than one race, several of the pTnX$_1$ RFLP types were shared among two or three races. Race 3 shared pTnX$_1$ RFLP types with races 1, 2, 4, and 5.

All of the Xoo DNA samples tested digested well with EcoRI. A continuous series of bands ranging in size from over 25 kb to less than one kb was observed after agarose gel electrophoresis of EcoRI-digested DNA. The DNA of all isolates tested from races 2, 4, 5 and 6 digested well with *Pst*I, but relatively few DNA fragments over six kb in size were observed. The remaining fragments formed distinctive patterns that characterized different strains. For races 1 and 3, two distinct lineages were detected using *Pst*I: those for which DNA samples could be digested with *Pst*I, and those for which DNA was not digestible with *Pst*I. RFLP analysis using pTnX$_1$ confirmed that the two lineages were distinct. The restriction enzyme *Xor*I, an isoschizomer of *Pst*I, was isolated from *Xanthomonas campestris* pv. *oryzae* (Wang et al., 1980). It is possible that some Xoo lineages modify their DNA to avoid digestion by an endogenous restriction system.

To determine whether RFLP data from analyses using pJEL101, pTnX₁, and *Pst*I provide consistent estimates of diversity, a comparison was made of the number of distinct RFLP types detected when the same set of 82 isolates was analyzed by all three methods. As shown in Table 1, the different methods showed different overall levels of diversity and different relative levels of diversity for the different races. Analysis of the set of of isolates by all three methods revealed substantially more diversity than was revealed by any one of the methods alone.

How can RFLP data on pathogen populations be useful? One application is in providing a basis for the selection of diverse tester strains for screening germplasm for resistance. We have used this approach to analyze the diversity of host and pathogen populations in a traditional agroecosystem in the Philippines. Race 5 is endemic to the mountainous Ifugao Province, an area where many traditional rice cultivars are grown. We selected strains representing different RFLP types from among race 5 strains, and tested them against a set of Ifugao traditional cultivars.

Diverse reactions were seen when over 300 traditional cultivars tested with three strains representing different RFLP types, and some of the cultivars showed differential reactions to the isolates. Further tests have been done on one of the cultivars, accession number 8106 of the International Rice Germplasm Center, confirming the differential reactions to the isolates. Differential reactions would not be expected unless the cultivars contain resistance genes not represented in the set of differential cultivars used to define the Xoo strains as a single race. It seems that information on the diversity of the pathogen may be very useful in screening germplasm to detect unrecognized resistance genes.

Table 2. Relative number of hybridizing bands for repetitive DNA elements in the genomes of *Pyricularia grisea* isolates infecting different hosts.

Isolate	Host Species	mtDNA genotype	Ralative Copy Number of Repeats					
			69/ *Bam*H1	612/ *Bam*H1	106/ *Bam*H1	46/ *Bam*H1	B21/ *Eco*RV	586/ *Eco*RI
R88107	*Oryza sativa*	A	+++++	+++++	+++++	+++++	+++++	+++++
R88374	*Oryza sativa*	A	+++++	+++++	+++++	+++++	+++++	+++++
V850261	*Oryza sativa*	A	+++++	+++++	+++++	+++++	+++++	+++++
JMB840328	*Oryza sativa*	A	+++++	+++++	+++++	+++++	+++++	+++++
Ec883	*Echinocloa colona*	A	+++++	+++++	+++++	+++++	+++++	+++
Lh88490	*Leersia hexandra*	A	+++++	+++++	· +++++	+++++	+++++	+++
Pd	*Paspalum distichum*	A	+++++	+++++	+++++	+++++	+++++	+++
Pr886	*Panicum repens*	A	+++++	+++++	+++++	+++++	+++++	+++
Lc7	*Leptochloa chinensis*	A	+++	+++++	+++++	+++	+++	+++
Bm88508	*Brachiaria mutica*	A	+++	+++	+++	+	+++++	+++
Ei88424	*Eleusine indica*	A	+	+	+++++	+++	+++	+++
Bd18	*Brachiaria distachya*	A	+	+	+++	+	+	+++
Re12	*Rottboellia exaltata*	A	0	0	+	+++	+	+++
Ce88454	*Cenchrus echinatus*	C	+++	+++	+++	+++++	0	+
Pp2	*Pennisetin purpureu*	C	+++	+++	+++	+++++	+++	+
Dc275	*Digitaria ciliaris*	F	+++	+++	+	+	+++++	+++
Er271	*Eragrostis sp*	F	+	+++	+	+	+++++	+++
Cr17	*Cyperus rotundus*	D	0	0	+	0	0	+
Cr88383	*Cyperus rotundus*	E	0	0	0	0	0	+++
Cb8959	*Cyperus brevifolios*	B	0	0	0	0	0	0

+++++ = more than 25 discernible bands
+++   = 10 - 20 discernible bands
+   = fewer than 10 discernible bands
0   = no discernible bands
⁺Probe probvided by DuPont

A subset of the Ifugao traditional cultivars were tested with 16 RFLP types from race 5. Individual plants were tested with multiple strains representing different RFLP types. Some of the cultivars appeared to be mixtures of resistant and susceptible genotypes. Accession 8106, for instance, consisted of a mixture of resistant, susceptible, moderately resistant, and differentially resistant individuals (Vera Cruz, Nelson and Mew, unpublished). Thus, the traditional rice cultivars grown in the Ifugao region are diverse with respect to bacterial blight resistance, and even a single cultivar may possess variability with respect to resistance. A diverse host population may provide minimal selection pressure for pests that overcome resistance, and may thus reduce the likelihood that virulent races become prevalent (Gould, 1988).

The bacterial blight pathogen shows considerable variation with respect to both virulence and molecular phenotypes. To understand the spatial scale of this variation, we recently conducted intensive sampling from fields cultivated under various conditions. We found that, in many cases, more than one RFLP type of the bacterium was present within a single field. In the future we hope to determine whether cultural practices, such as those affecting the diversity of the rice crop, can affect the diversity and evolution of the pathogen. If so, then cultural practices may be adopted that maximize the durability of resistance to bacterial blight of rice.

RICE BLAST FUNGUS

Rice blast disease, caused by *Pyricularia grisea* Cavara (teleomorph *Magnaporthe grisea* [Hebert] Barr comb. nov; Rossman et al., 1990), is the most important fungal disease of rice. The fungus is considered to be highly variable. Hundreds of pathogenic races have been identified based on their virulence spectra on rice cultivars carrying different resistance genes (Ou, 1985). Resistance to rice blast disease is notoriously short-lived in the field.

The objectives of our research on the blast fungus have been:

a. To determine the extent of gene flow between weed- and rice-infecting populations of the pathogen;

b. To evaluate the variability among rice-infecting strains of *P. grisea* in the Philippines;

c. To evaluate the extent of association between virulence phenotype and molecular phenotype using a repetitive DNA probe; and

d. To determine the extent to which the pathogen diversity present at screening sites used to assess the resistance of rice breeding materials reflects the diversity of the pathogen in farmers' fields (study in progress).

P. grisea infects more than 50 gramineous hosts (Ou, 1980). Although weed-infecting strains generally do not infect rice in cross-inoculation experiments, it is possible that *P. grisea* populations growing on weeds could infect rice at low frequency, perhaps providing a source of new races of the pathogen. Hamer et al. (1990) used a class of repetitive DNA sequences to probe DNA samples from several blast fungus strains from rice and non-rice hosts from various countries. These repetitive DNA elements, termed MGR sequences, were abundant in rice-infecting isolates, but not highly conserved in the non-rice infecting strains tested.

Table 3. Frequencies and geographic distribution of different RFLP types detected by probe POR 613 in rice-infecting isolates of *P. oryzae* in the Philippines.

RFLP type	Frequency (No. of Provinces)	Distribution	Virulence Clusters*
a	0.14	11	I, III, IV, IX
b	0.1	3	V, VIII, III, VII
c	0.1	1	V, VII, VIII
d	0.09	5	III, IV, II
e	0.06	5	II, IV, VI, VIII
f	0.04	2	II, VI
others	0.01-03		

*Isolates were grouped into 12 clusters based on reactions to 20 rice cultivars.

Because only a single Philippine strain of *P. grisea* from a non-rice host was tested in the study of Hamer et al. (1990), we decided to extend the analysis of gene flow between rice-and weed-infecting isolates in the Philippines. We surveyed a larger number of isolates collected from various weeds that commonly surround rice fields, and tested for RFLPs among the fungal mitochondrial DNAs (mtDNAs) and among nuclear DNAs using a set of repetitive DNA probes (Table 2). While some of the weed-infecting isolates showed mtDNA restriction digest patterns that were distinct from those of the rice-infecting isolates, other weed-infecting isolates showed exactly the same mtDNA pattern as were seen for the rice-infecting isolates. This suggested that *P. grisea* populations infecting some weeds are more closely related to rice-infecting populations than are populations infecting other weed hosts.

To examine the extent of this relatedness, a set of repetitive DNA elements isolated from the fungal genome were used as probes against restriction digests of genomic DNA from the fungal isolates (Table 2). Five of these probes, isolated at IRRI (Borromeo, Nelson and Leung, unpublished), showed polymorphisms between the weed- and rice- infecting isolates, and the number of bands showing hybridization with the five probes differed between the rice-infecting strains and some of the weed-infecting strains. Using the MGR586 probe (provided by Valent and Chumley of DuPont; Hamer et al., 1990) all the weed-infecting isolates could be clearly distinguished from the rice-infecting isolates on the basis of the number of hybridizing bands. It appears, then, that repetitive nuclear DNA markers may evolve faster than mtDNA. Based on the repetitive nuclear DNA markers, rice-and non-rice infecting populations of the fungus are quite distinct, at least among the populations sampled to date. Thus our data suggest that, for the populations examined in our study and that of Hamer et al. (1990), we need not urge farmers to remove blast-infected weeds growing near their rice-fields.

In developing resistant rice lines, breeders test their materials at screening sites that are disease "hot-spots". Implicit in this approach is the assumption that the diversity of the population present at the screening site adequately represents the diversity of the pathogen in farmers' fields. To evaluate this assumption, we are currently comparing the diversity of *P. grisea* isolates collected from farmers' fields in the Philippines with the diversity of the pathogen population present at a key screening site in the Philippines.

We have conducted RFLP analysis of a collection of rice-infecting isolates from various sites in the Philippines. Using a repetitive DNA probe designated pPGR613, a high degree of variability was detected. Lineages of the fungus, as defined by RFLP types, varied in geographical distribution and in the frequency with which they were represented in the collection (Table 3). For instance, one RFLP type was limited to a single province, whereas another was found distributed among 10 provinces.

For the population examined, the RFLP types detected by probe pPGR613 were not strongly associated with pathogenic phenotype, as defined by testing on a set of differential rice cultivars. That is, single RFLP types were detected among multiple pathotypes, and single pathotypes were associated with multiple RFLP types. In contrast, Levy et al. (1991) found that among a collection of North American strains of the blast fungus, RFLP types defined by probe MGR 586 were strongly associated with pathotype.

CONCLUSIONS

DNA probes are useful markers for studying pathogen population genetics. Repetitive probes may produce characteristic "fingerprints" that identify strains or lineages of asexually-reproducing pathogens. Relatedness among strains can be inferred, giving insight into the processes of race evolution, gene flow, and other aspects of pathogen population biology. Information about pathogen population dynamics may be useful in designing improved strategies for disease management. Knowledge of pathogen lineages can also be used in the selection of diverse tester strains used in screening of breeding material for crop improvement.

Different probes may reveal somewhat different information about a population. The specific estimates of diversity obtained for a given population depend on the specific probe employed in the analysis. Probes may vary in the extent to which they reveal RFLP patterns that are associated with pathogenic race. If there is a perfect association between RFLP type and race, a probe may be used to identify race. The extent of association between identified molecular and pathogenic phenotypes depends upon the characteristics of the population tested, of the differential cultivars chosen for virulence analysis, and of the probe and enzyme used for the analysis of RFLP type.

ACKNOWLEDGEMENTS

The work discussed here was conducted at the International Rice Research Institute (IRRI), in collaboration with Kansas State University, Washington State University, and the University of Wisconsin. The collaborators include: A. Ardales[1], M. Baraoidan[1], J.M. Bonman[1], E. Borromeo[1], J.E. Leach[2], H.Leung[1,3], T.W. Mew[1], R. Nelson[1], A.K. Raymundo[1,4], and C.M. Vera Cruz[1]. The research was supported by IRRI and the Rockefeller Foundation's Rice Biotechnology Program.

[1]Division of Plant Pathology, IRRI, P.O. Box 933, Manila, Philippines.

[2]Dept. of Plant Pathology, Kansas State University, Pullman, WA, 99164-6430, U.S.A.

[4]Inst. of Biological Sciences, University of the Philippines at Los Banos, College, Laguna, Philippines

REFERENCES

Ezuka, A. and Horino, O. 1974. Classification of rice varieties and *Xanthomonas oryzae* strains on the basis of their differential reactions. Bulletin Tokai-Kinki National Agricultural Experimental Station 27: 1-19.

Ezuka, A. and Sakaguchi, S. 1978. Host-parasite relationship in bacterial leaf blight of rice caused by *Xanthomonas oryzae*. Rev. Plant Prot. Res. 11: 93-118.

Gould, F. 1988. Evolutionary biology and genetically engineered crops-Consideration of evolutionary theory can aid in crop design. BioScience 38 (1) :26-33.

Hamer, J.E., Farrel, L., Orbach, M.J., Valent, B., and Chumley, F. 1989. Host species-specific conservation of a family of repeated DNA sequences in the genome of a fungal plant pathogen. PNAS 88: 9981-9985.

Kearney, B. and Staskawicz, B.J. 1990. Characterization of IS476 and its role in bacterial spot disease of tomato and pepper. J. Bacteriol. 172:143-148.

Leach, J.E., Rhoads, M.L., Vera Cruz, C.M., White, F.F., Mew, T.W. and Leung, H. In preparation. Genetic diversity and population structure of the bacterial blight pathogen of rice as revealed by virulence and DNA polymorphism analysis.

Leach, J.E., White, F.F., Rhoads, M.L., and Leung, H. 1990. A repetitive DNA sequence differentiates *Xanthomonas campestris* pv. *oryzae* from other pathovars of *X. campestris*. Molecular Plant- Microbe Interactions 3(4):238-246.

Lee and Cho, S.Y. 1990. Variation in races of rice blast fungus. (Paper presented to the Annual Meeting of the American Phytopathological Society, August 4-8, 1990, Grand Rapids, Michigan; abstract to be published in *Phytopathology*.

Levy et al., 1991 : Levy, M., Romao, J., Marchetti, M.A. and Hamer, J.E. 1991. DNA fingerprinting with a dispersed repeated sequence resolves pathotype diversity in the rice blast fungus. The Plant Cell 3: 95-102.

McDonald, B.A., McDermott, J.M., Goodwin, S.B. and Allard, R.W. 1989. The population biology of host-pathogen interactions. Annu. Rev. Phytopathol. 27: 77-94.

Mew, T.W., 1987. Current status and future prospects of research on bacterial blight of rice. Ann. Rev. Phytopathol. 359-382.

Mew T.W. and Vera Cruz. 1985. Virulence of *Xanthomonas campestris* pv. *oryzae* in the Philippines. Phytopathology 75: 1316. (Abstract)

Ou, S. H. 1980. Pathogen variability and host resistance of the rice blast fungus, *Pyricula ria oryzae* Cavara. Ann. Rev. Phytopathol. 18: 167-187.

Ou, S.H. 1985. Rice Diseases (Commonwealth Mycological Institute, Kew, Surrey, U.K.), p. 109.

Parlevleit, J.E. 1983. Durable resistance in self-fertilizing annuals. In Durable Resistance in Crops. F. Lamberti, J.M. Walker, and N.A. Van der Graaff, Eds. Plenum Publishing Co.

Parlevleit, J.E. 1983. Durable resistance in self-fertilizing annuals. In Durable Resistance in Crops. F. Lamberti, J.M. Walker, and N.A.. Van der Graaff, Eds. Plenum Publishing Co.

Raymundo, A.K., Nelson, R.J., Ardales, E.Y., Baraoidan, M.R. and Mew, T.W. 1990. A simple method for detecting genetic variation in *Xanthomonas campestris* pv. *oryzae* (Xco) by restriction fragment length polymorphism (RFLP). International Rice Research Newsletter 15 (4): 8-9.

Rossman, A., Howard, R.J., and Valent, B. 1990. *Pyricularia grisea*, the correct name for the rice blast disease fungus. Mycologia 32 (4) : 509-512.

Swings, J., Vand den Mooter, M., Vauterin, L., Hoste, B., Gillis, M., Mew, T.W., and Kersters, K. 1990. Reclassification of the causal agents of bacterial blight (*Xanthomonas campestris* pv. *oryzae*) and bacterial blight (*Xanthomonas campestris* pv. *oryzicola*) of rice as pathovars of *Xanthomonas oryzae* (ex Ishiyama 1922) sp. nov., nom. rev. Internat. J. System. Bacteriol. 40: 309-311.

Wang, R.Y. -H., Shedlarski, J.G., Farber, M.B., Kuebbing, D.,and Ehrlich, M. 1980. Two sequence-specific endonucleases from X*anthomonas oryzae*: characterization and unusual properties. Biochimica et Biophysica Acta 606: 371-385.

ENVIRONMENTAL CONTROL OF MICROBIAL GENE EXPRESSION AND EVOLUTION

A.M. Chakrabarty

Dept. of Microbiology & Immunology
University of Illinois College of Medicine
P.O. Box 6998, Chicago, Illinois 60680
U.S.A.

INTRODUCTION

Microorganisms live in an ever-changing environment. Human activities contribute significantly to these environmental changes, particularly because microorganisms not only use the external environment as their habitat, but also the human body. Thus drugs and antibiotics are often given to human patients to eradicate a particular microbial infection, or new chemicals are introduced into the environment by the chemical industry in the form of herbicides/pesticides, refrigerants, solvents, etc. that sometimes can create toxicity problems and at other times can become a major source of carbon available to the microorganisms. Since microorganisms must express particular sets of genes to survive in a highly stressed environment, or evolve new genetic capabilities to be able to break down new types of synthetic compounds, they must develop appropriate and rapid response to such changes in their environment to be able to cope with such changes.

Bacteria constantly sense the external environment and modulate their physiological activity through activation or repression of gene expression. In the most widely studied cases, microorganisms employ a two component system where one component, usually a trans-membrane regulatory protein, senses the environment, binds with a particular ligand, and modifies a second component, usually a DNA binding regulatory protein, which in its modified form is able to activate a specific promoter to allow gene expression to occur. Such a signal transduction system has been demonstrated to be functional in a large number of cases such as chemotaxis, nitrogen or phosphate regulation, osmoregulation, oxygen and metabolic activity, sporulation, etc. (1,2). Similarly, starvation for a carbon source, which allows enhanced cyclic AMP synthesis, is known to induce expression of a set of genes in *Escherichia coli* mediated sometimes, but not in all cases, by cAMP-CRP complex (3). Such environmental control of gene expression is also exhibited by pathogenic microorganisms during infection of mammalian cells. Thus the expression of *Salmonella* adherence and invasiveness to enter host cells is dependent on the growth conditions of the cells, particularly the limitation of oxygen may be the environmental cue that triggers such expression (4). Similarly, the transcriptional initiation from bvg P1 promoter for bvgABC operon in *Bordetella pertussis* is known to be responsive to environmental factors (5).

Biotechnology and Environmental Science: Molecular Approaches
Edited by S. Mongkolsuk et al., Plenum Press, New York, 1992

99

Another factor that affects microbial activity significantly in the environment is the release of highly chlorinated synthetic chemicals. Such chemicals are widely used, and because they are new to the environment, they often remain unattacked by natural micro-organisms, leading to their persistence. In many cases, such persistence has led to problems of toxicity in human and animal populations.

Many chlorinated compounds can pose problems of toxicity for microorganisms, but can also act as major source of their carbon and energy. Thus, they often evoke a ready response from natural microorganisms leading to their mineralization and detoxification. In this short article, I discuss two examples where the environmental factors exert profound influence in triggering gene expression or in accelerating the process of evolution of new genes in natural microorganisms. The consequence of the former response leads to pathogenicity of the organisms, while the results of the latter response allow complete mineralization of some synthetic, chlorinated compounds leading to their detoxification.

I. MOLECULAR ECOLOGY OF MUCOID PSEUDOMONAS AERUGINOSA IN CYSTIC FIBROSIS LUNG

An interesting example where environmental factors govern the expression of a set of genes leading to a major medical problem is the case of *Pseudomonas aeruginosa* infection in the lungs of cystic fibrosis (CF) patients. Cystic fibrosis is a genetic disease prevalent among the Caucasian population. The CF patients usually accumulate in their lungs highly viscous abnormal fluids, which are quite sticky and dehydrated and similar to other CF glands contain appreciable amounts of Na^+, K^+ and Cl^- ions (6). Accumulation of the abnormal fluids triggers bacterial infections. The initial infections are usually due to *Haemophilus* or *Staphylococcus*, but they can be eliminated by antibiotic treatment. Subsequent infections are usually due to *P. aeruginosa*. The initial infecting *P. aeruginosa* cells are nonmucoid which increasingly turn mucoid due to production of an exopolysaccharide capsule called alginic acid. Alginate is a linear copolymer consisting of b-1, 4-linked D-mannuronic acid and variable amounts of its C-5 epimer L-guluronic acid. It is important to note that *P. aeruginosa* rarely produces alginate outside the CF lung. Apparently, alginate forms a strong gel around the *P. aeruginosa* microcolonies in the upper respiratory tract of the CF patients which protects such cells from subsequent treatments with antibiotics or from phagocytosis. The mucoidy is an unstable character and mucoid *P. aeruginosa* cells revert to nonmucoidy at a rapid rate outside the CF lung. The excessive amount of the mucus produced by mucoid *P. aeruginosa* (Fig. 1) in the CF lung adds to the clogging of the lung in addition to production of elastase and other virulence factors that damage the lung tissue, leading to its failure and usually the death of the patient.

Why are alginate genes, that are normally not expressed in any other environment, specifically expressed in the CF lung? In order to understand the nature of the enzymes involved in the biosynthesis of this capsular polysaccharide and their regulation, we isolated a stable alginate-positive (Alg⁺) strain of *P. aeruginosa* (strain 8830) from the wild type unstable CF isolate strain 8821, derived a large number of Alg⁻ mutants by EMS mutagenesis, cloned the alg genes by complementation of such Alg⁻ mutants and mapped them on the chromosome (7). The arrangement on two different parts of the *P. aeruginosa* chromosome (34 min and 10 min) of a cluster of genes (8) specifying a number of biosynthetic enzymes (34 min) or regulatory proteins (10 min) is shown in Fig. 2B.

Subcloning various fragments of either cluster and hyper-expressing them under the tac promoter has allowed identification (9, 10) of not only the biosynthetic enzymes PMI-GMP and GMD (Fig. 2A) but also the regulatory proteins AlgR1, AlgR2 and AlgR3 that control the

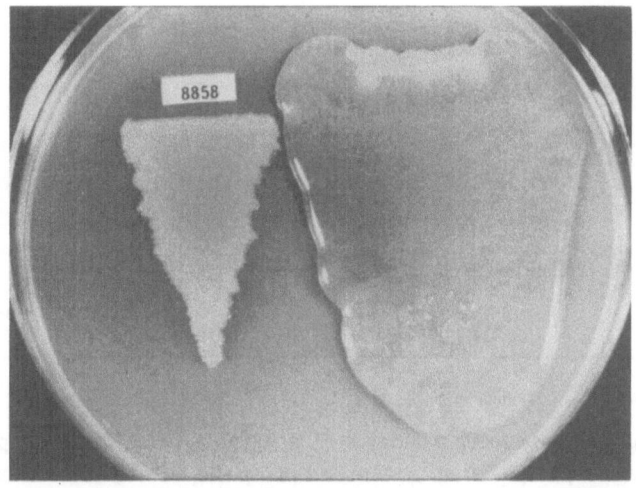

Figure 1. Morphological features of a nonmucoid P. aeruginosa 8858 and its mucoid variant.

expression of biosynthetic genes (11, 12, 13). Recently, the gene for the second enzyme of the pathway (Fig. 2A), PMM, has been completely sequenced and the PMM enzyme has been characterized as a 50 KD protein (13a). Interestingly, the gene for PMM is not located within the cluster shown in Fig. 2B.

The arrangement of the alg genes in Fig. 2B demonstrates the presence of the gene algD, encoding GMD, at one extreme of the cluster which has at the other end the algA gene which encodes a bifunctional 53 KD protein PMI-GMP (Fig. 2A). Both algA (14) and algD (15) have been completely sequenced. The transcription of the alg genes located in this cluster has been shown to proceed from right to left (8), as shown by arrows. Several of the alg genes between algA and algD have also been sequenced and a putative promoter has been located upstream of the alg76 gene (Chu, L., May, T. and Misra, T.K., unpublished observations). The enzymes PMI-GMP and GMD have also been purified and partly characterized (15a, 16). The expression of the algD gene, which shows a number of direct and inverted repeats at its upstream promoter region (15), is normally silent and the promoter must be activated under specific environmental conditions (17) for expression of the algD gene. In nonmucoid P. aeruginosa cells, the algD promoter is not activated, while in mucoid cells, the promoter is activated to allow production of alginate (10). As mentioned previously, at least 3 regulatory proteins, AlgR1, AlgR2, and AlgR3, have been shown to be involved in the activation process. AlgR1 is a member of the two component sensory transduction system, having homology with NtrC and OmpR (11), while AlgR2 (12) and AlgR3 are not part of the two component system. AlgR3 is a highly basic protein with homology to sea urchin histone H1 protein (13). AlgR1 has recently been purified and has been shown to be a DNA binding protein having the ability to bind with more than one site in the upstream region of algD (17a).

The emergence and proliferation of mucoid P. aeruginosa cells primarily in the lungs of the CF patients, and not in the environment, suggests that such cells have an ecological niche based on the unique ecology of the CF lung. How does the environment of the CF lung control the induction of mucoidy in P. aeruginosa?

Although the details are not clear, a picture is emerging that points out to the key

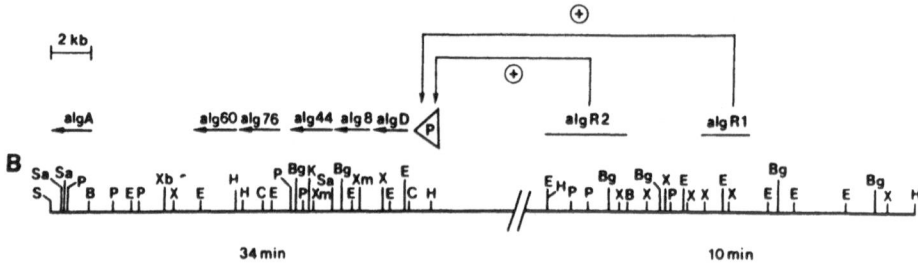

Figure 2. A: Alginate biosynthetic pathway in P. aeruginosa. F6P, fructose-6-phosphate; M6P, mannose-6-phosphate; MIP, mannose-1-phosphate; GDPM, GDP-mannose; GMA, GDP-mannuronic acid; PMI, phosphomannose isomerase; PMM, phosphomannomutase; GMP, GDP-mannose pyrophosphorylase; GMD, GDP-mannose dehydrogenase.

B: Restriction map of the alginate structural gene cluster and the alginate regulatory region, located at 34 min and 10 min respectively, on the P. aeruginosa chromosome. The algD promoter is marked with a P. Abbreviations, S, SmaI; Sa, SalI; P, PstI; B, BamHI; E, EcoRI; Xb, XbaI; X, XhoI; H, HindIII; C, ClaI; Bg, BglII; K, KpnI; Xm, XmaI.

environmental factors in the CF lung contributing to the transition to mucoidy. Berry et al (18) and DeVault et al (19) have demonstrated that either high osmolarity (high NaCl concentration) or ethanol (a commonly used dehydrating agent) will allow activation of the algD promoter (Table 1). Activation of the promoter in presence of high NaCl concentration does not, however, lead to alginate production (18), while ethanol-induced activation, during the late stage of the growth when the cells reach starvation phase, will allow genotypic switching to mucoidy resulting in alginate production (19). This suggests that a number of environmental signals might operate simultaneously in order to allow activation of the algD promoter to a level where it results in a genotypic switching. The activation of the algD promoter requires participation of a number of auxiliary proteins such as DNA gyrase (19) and the alternate sigma factor s54 or RpoN (20). Exactly how the various regulatory proteins interact to effect algD promoter activation,whether negative super-coiling or looping of the upstream DNA (since AlgR1 appears to bind with a DNA region several hundred bp upstream of the transcription initiation site) is required to allow contact with the RpoN component of the RNA polymerase for transcription and whether a particular level of activation of the algD promoter turns on the switch to mucoidy, are questions that remain unanswered at present. What is interesting is that two environmental factors, high salts and ethanol, which is a membrane-perturbing agent producing effects similar to dehydration, are two unique characteristics of the CF mucus environment that trigger mucoidy in P. aeruginosa, presumably by activating the algD promoter.

II. EVOLUTION OF DEGRADATIVE GENES IN RESPONSE TO RELEASE OF SYNTHETIC CHLORINATED COMPOUNDS

In the preceding section, I discussed how environmental factors modulate gene expression in bacteria to enable them to establish themselves in a particular ecological niche, or to exert pathogenic characteristics detrimental to human health. Microorganisms are, however, well known for their nutritional adaptability and are widely used in bioremediation or municipal sludge treatment. A problem of major magnitude confronting the industrialized countries is the pollution of the environment due to release of highly chlorinated compounds manufactured by the chemical industry. Such compounds are usually very persistent, because natural microorganisms have not developed the capability to degrade them to any significant extent. Compounds such as polychlorinated biphenyls (PCBs), highly chlorinated dioxins, chlorophenols etc. have created major toxicological problems due to their recalcitrance to microbial attack.

An approach to addressing the problem of environmental pollution due to the release of highly chlorinated toxic chemicals would be to accelerate the process of evolution of the degradative genes in natural microorganisms against such compounds. Since many of the chlorinated compounds have been used as herbicides and pesticides and therefore released into the environment in massive amounts for the last several decades, it is pertinent to ask to what extent natural microorganisms might have evolved the genes encoding degradation of some of these compounds. Work from a number of laboratories (21) has demonstrated that chlorinated compounds such as 3-chlorobenzoic acid or 2,4-dichlorophenoxyacetic acid (2,4-D) are readily utilized by natural microorganisms. In both the cases, natural microoir ganisms have evolved the 3-chlorocatechol (3-Clc) and the 2,4-D degradative genes in the form of plasmids (22). An important question in this regard is that when microorganisms evolve new biodegradative genes in response to a new chemical, how are such genes recruited and organized into a well-regulated pathway?

To determine how new biodegradative genes evolve in nature , we have analyzed the chlorocatechol pathway encoded by the plasmid pAC27 and compared such genes with those

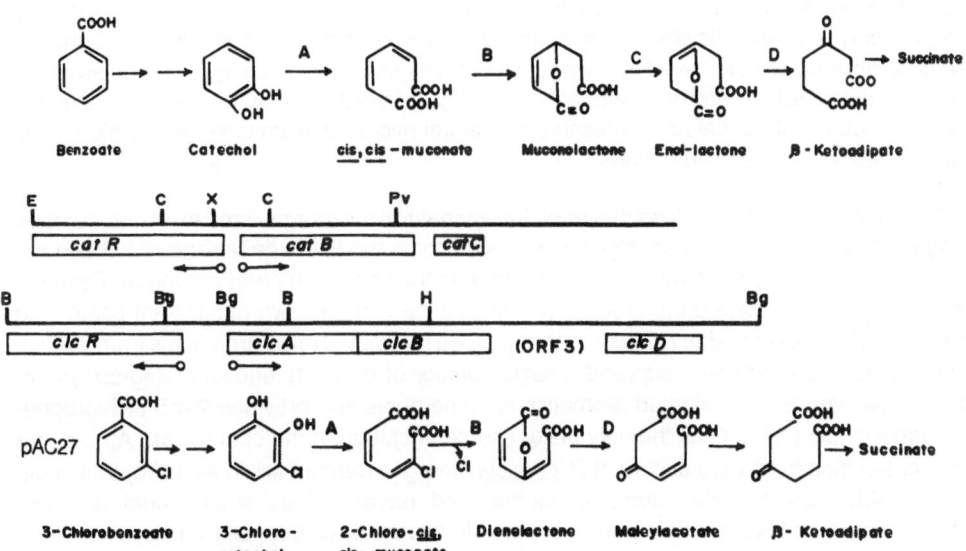

Figure 3. Catechol and chlorocatechol metabolic pathways and organization of the cat and clc operons. Note the divergent transcription of the catR gene from the catBC operon and that of the clcR gene from the clcABD operon.

encoding degradation of the natural compound catechol. There are 3 genes termed clcA, clcB and clcD borne on the plasmid pAC27 that allow conversion of 3-chlorocatechol to maleylacetic acid which is further metabolized through the b-ketoadipate pathway (22). The three clc genes have been sequenced and found to be clustered as clcABD operon (23) under a single promoter (Fig. 3). The clcABD operon is under positive control by a regulatory gene clcR which is transcribed divergently from the operon. The catechol degradative genes cat B and catC are similarly organized as an operon (24) under a single promoter (Fig. 3). The catBC operon is also under positive control by a regulatory gene catR, which is transcribed divergently from the catBC operon. Thus the organization of the catBC operon/catR gene is remarkably similar to that of the clcABD operon/clcR gene (Fig. 3).

In addition to the similarity of the genetic organization, the individual genes in these two operons also show appreciable homology. For example, catB gene has extensive homology with the clcB gene. The catR gene has recently been sequenced (25), and the sequence of the clcR gene has also been determined (Coco, W.M., unpublished observations). The catR and clcR gene products show 33% identity (26). Such homologies among structural and regulatory genes of the clc operon with those of the cat operon, coupled with the similarity of their organizations, appear to suggest that the plasmid-borne clc genes were recruited from the chromosomal cat genes.

The plasmid pJP4, which encodes the 2,4-D pathway (Fig. 4), allows conversion of 2,4-D to 3,5-dichlorocatechol via 2,4-dichlorophenol (27). The tfdC, tfdD and tfdE genes which allow conversion of 3,5-dichlorocatechol to chloromaleylacetic acid (Fig. 4) show significant homology to the corresponding clcA, clcB and clcD genes of the pAC27 plasmid both by Southern hybridization (28) and by nucleotide sequence analysis (29). It is interesting to note that the pJP4-borne clc genes, which are part of a 2,4-D degradative gene cluster, show such significant homology to the pAC27-clc gene cluster, even though the pAC27 and pJP4 are completely different plasmids found in two different bacterial genera which were isolated in two different countries (pAC27 in P. putida from U.S.; pJP4 in Alcaligenes eutrophus isolated in Australia). It thus appears that natural evolution for genes encoding degradation of a synthetic chlorinated compound involved recruitment of genes that encode an analogous function followed by divergence of such genes so that the gene products would have an extended or different substrate specificity. Evolution of such genes presumably occurs because of the selection pressure on natural microorganisms exerted by high, toxic concentrations of such compounds.

It should be emphasized that even though natural microorganisms exist that degrade simple chlorinated compounds, highly chlorinated compounds are quite recalcitrant and are not known to be degraded by natural microflora as their sole source of carbon and energy. Is it possible to accelerate the process of evolution so that a known recalcitrant compound can be made biodegradable? Towards this goal, we inoculated microorganisms from various dump sites in a chemostat, supplied a large number of plasmids encoding degradation of chlorinated and non-chlorinated aromatics as gene pools and provided 2,4,5-trichlorophenoxyacetic acid (2,4,5-T) as the only major source of carbon to the chemostat. After about nine or ten months, a pure culture of P. cepacia emerged (termed strain AC1100) that could utilize 2,4,5-T as its sole source of carbon and energy. This strain could not only utilize 2,4,5-T or 2,4,5-trichlorophenol as its sole source of carbon and energy, but could additionally dechlorinate a large number of chlorophenols (21). This process of generating bacterial strains that have developed the capability to utilize a normally recalcitrant compound has been termed directed evolution (26). How do bacteria under a stressed environment in the chemostat evolve the genes for the degradation of the target chemical? In order to have an insight in the directed evolution of 2,4,5-T degradative (tft) genes, we have isolated both

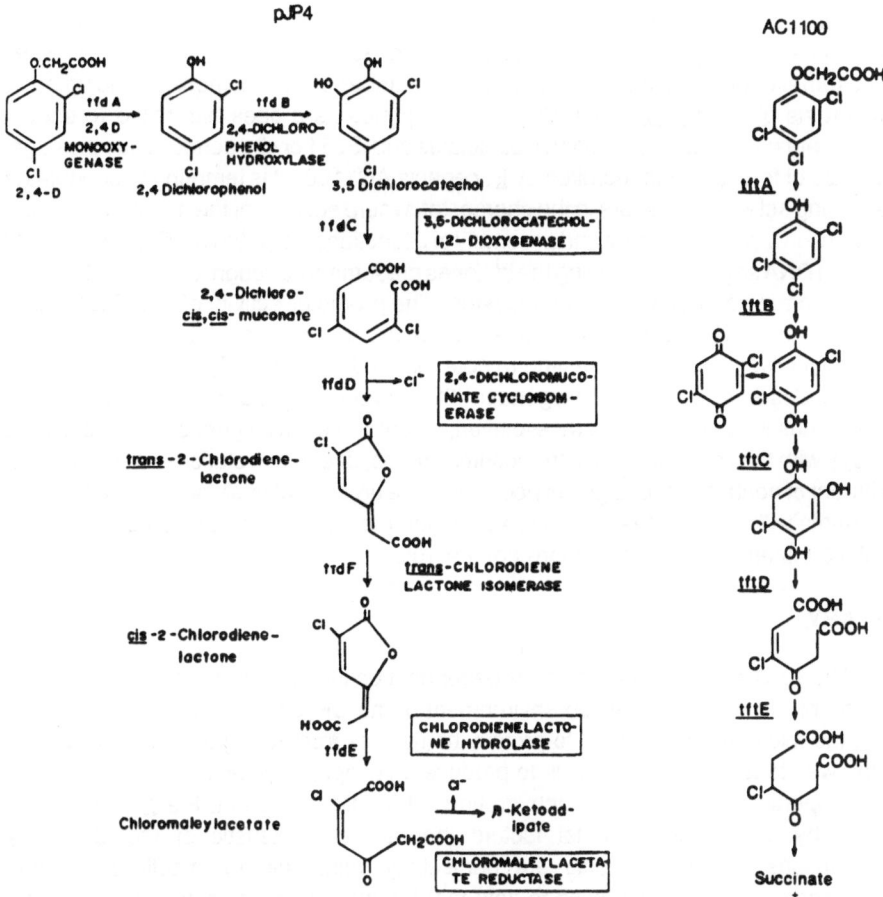

Figure 4. Metabolic pathways of dissimilation of 2,4-D encoded by pJP4 plasmid and a tentative pathway for 2,4,5-T encoded by the chromosomal genes of P. cepacia strain AC1100.

spontaneous and Tn5 induced mutants in the Tft pathway (Fig. 4), and cloned a number of genes that complement such Tft⁻ mutants to Tft⁺. For example, a mutant PT88 accumulates an intermediate CHQ (5-chloro, 2-hydroxyhydroquinone) from 2,4,5-T because it lacks the CHQ oxygenase enzyme (30). A 4 kb BamHI-PstI fragment from AC1100 has been cloned that complements mutant PT88 to Tft⁺. This gene has been termed chq (30). A repeated sequence, initially termed RS1100, has also been characterized in the genome of AC1100, which is present with a minimum copy number of 30. This repeated sequence has been completely sequenced which shows the presence of characteristic long terminal inverted repeats, duplications of the target sites at the site of insertion, and the presence of a transposase type of gene (31). Indeed Haugland et al. (32) have recently demonstrated the transposability of RS1100, which has therefore been designated as IS931. IS931 has two interesting characteristics: it promotes gene expression at the site of insertion and it occasionally carries intervening DNA during transposition. Another interesting feature of IS931 is that even though there are more than 30 copies of IS931 on the genome of P. cepacia AC1100, most of the copies are restricted to a small region (about one third) of the chromosome of AC1100 near the tft genes. Thus many Tft⁻ spontaneous segregants lose large numbers of copies of IS931 and the tft genes without showing loss of essential chromosomal genes. Colony hybridization of a number of pseudomonads, including laboratory strains of P. cepacia, with IS931 or chq genes as probes has failed to show any homology between natural pseudomonad isolates and IS931 or chq genes. Thus these genes seem to be of foreign origin recruited by P. cepacia AC1100. It is tempting to speculate that under strong selective pressure in the chemostat, where survival of the microorganisms was dependent on rapid recruitment and assembly of a functional Tft pathway, P. cepacia AC1100 recruited a DNA segment containing the tft genes presumably through a transpositional event by IS931 from a non-pseudomonad ancestor. The promoter activity of the IS931, inserted in the upstream regions of tft genes, ensured the expression of these foreign genes in P. cepacia AC1100. It is interesting to note that both the chq (30, 31) and the tmo gene, which encodes 2, 4, 5-T monooxygenase (33), harbor copies of IS931 in their upstream regions. Thus in contrast to natural evolution, where the evolved genes show considerable homology with the corresponding chromosomal genes, directed evolution of genes may allow recruitment of nonhomologous genes because of the urgency of evolving such genes. Future experiments with other cases of directed evolution may demonstrate whether such a mode of evolution is an exception or a more general rule.

SUMMARY

The survival of microorganisms is dependent on their ability to respond to a changing environment. In the very stressed environment of the CF lung, with salty and dehydrated mucus, the microorganisms need to protect themselves from losing their intracellular water and one way to accomplishing this is to produce an exopolysaccharide capsule with strong gelling properties as a barrier to dehydration. It is interesting that the algD promoter is activated by those environmental factors that are characteristic of the CF disease, which explains why CF patients are particularly vulnerable to infections by mucoid P. aeruginosa. It is also interesting to note that the alginate capsule, which is presumably produced to protect P. aeruginosa from intracellular dehydration, also affords protection

Table 1. Activation of the algD promoter by environmental factors.

Environmental Factor	Concentration	C230 activity (mU/mg protein)	
		PAO1 (pVD2X) nonmucoid	8821 (pVD2X) mucoid
NaCl	0	1,129	7,654
	0.2 M	1,653	26,364
	0.35 M	3,302	31,255
Ethanol	0%	1,129	7,413
	0.5%	1,937	19,887
	1.0%	13,716	24,313

The activation of the algD promoter has been measured by measuring catechol 2,3-dioxygenase (C230) activity, since a construct containing the xylE gene placed under the algD promoter was used in these experiments (see ref. 18 and 19). One unit of catechol 2,3-dioxygenase activity is defined as the amount of enzyme oxidizing one mmol of catechol to 2-hydroxymuconic semialdehyde, a product with a molar extinction coefficient of 4.4×10^4 at 375 nm.

against antibiotics and antibodies to the detriment of the human patients. When the environment is beset with chlorinated compounds, the immediate response of natural microorganisms is to evolve degradative genes in the form of a plasmid. As a first step, they tend to recruit genes that allow degradation of a structurally analogous non-chlorinated compound, which undergo mutational or recombinational divergence to the appropriate genes whose products have broad substrate specificities to include chlorinated compounds. Such evolutionary processes apparently are not very effective for highly chlorinated compounds for which appropriate enzyme systems have not yet fully developed. In a single incidence of directed evolution, the chromosomal DNA shows the presence of multiple copies of a transposable element near the evolved genes, suggesting that an accelerated process of evolution may bypass the requirement of genetic relatedness of the evolved genes to the genome of the recruiting cells.

ACKNOWLEDGEMENT

The molecular ecology of mucoid P. aeruginosa in CF lung part of the work is funded by Public Health Service grant AI 16790-12 from the National Institute of Health and a grant Z061-9-2 from the Cystic Fibrosis Foundation. The evolution of degradative genes part is supported by a Public Health Service grant ES 04050 from the National Institute of Environmental Health Sciences and by a grant DMB-87 21743 from the National Science Foundation. I thank all my graduate students and post-doctoral colleagues, present and past, whose dedication and hard work made this work possible.

REFERENCES

1. Ronson, C.W., Nixon, B.T. and Ausubel, F.M. (1987). Cell 49:579-581.
2. Stock, J.B., Ninfa, A.J. and Stock, A.M. (1989). Microbiol. Rev. 53:450-490.
3. Matin, A., Auger, E.A., Blum, P.H. and Schultz, J.E. (1989). Ann. Rev. Microbiol. 43:293-316.
4. Lee, C.A. and Falkow, S. (1990). Proc. Natl. Acad. Sci. USA 87:4304-4308.
5. Roy, C.R., Miller, .J.F. and Flakow, S. (1990). Proc. Natl. Acad. Sci. USA 87:3763-3767.
6. McPherson, M.A. and Goodchild, M.C. (1988). Clin. Sci. 74:337-345.
7. Darzins, A., Wang, S.K., Vanags, R.I. and Chakrabarty, A.M. (1985). J. Bacteriol. 164:516-524.
8. Wang, S.K., Sa-Correia, I., Darzins, A. and Chakrabarty, A.M. (1987). J. Gen. Microbiol. 133:2303-2317.
9. Darzins, A., Nixon, L.L., Vanags, R.I. and Chakrabarty, A.M. (1985). J. Bacteriol. 161:249-257.
10. Deretic, V., Gill, J.F. and Chakrabarty, A.M. (1987). J. Bacteriol. 169:351-358.
11. Deretic, V., Dikshit, R.,Konyecsni, W.M., Chakrabarty, A.M. and Misra, T.K. (1989). J. Bacteriol. 171:1278-1283.
12. Kato, J., Chu, L., Kitano, K., DeVault, J.D., Kimbara, K., Chakrabarty, A.M. and Misra, T.K. (1989). Gene 84:31-38.
13. Kato, J., Misra, T.K. and Chakrabarty, A.M. (1990). Proc.Natl.Acad. Sci. USA 87:2887-2891.
13a.Zielinski, N.A., Chakrabarty, A.M. and Berry, A. (1991). J. Biol. Chem. 266: 9754-9763.
14. Darzins, A., Frantz, B., Vanags, R.I. and Chakrabarty, A.M. (1986). Gene 42:293-302.
15. Deretic, V., Gill, J.F. and Chakrabarty, A.M. (1987). Nucleic Acid Res. 15:4567-4581.
15a.Shinabarger, D., Berry, A., May, T.B., Rothmel, R., Fialho, A. and Charkrabaty, A.M. (1991). J. Biol. Chem. 266: 2080-2088.
16. Roychoudhury, S., May, T.B., Gill J.F., Singh, S.K., Feingold, D.S. and Chakrabarty, A.M. (1989). J. Biol. Chem. 264:9380-9385.

17. DeVault, J.D., Berry, A., Misra, T.K., Darzins, A. and Chakrabarty, A.M. (1989). Bio/ Technology $\underline{7}$:352-357.

17a. Kato, J. and Chakrabarty, A.M. (1991). Proc. Natl. Acad. Sci. USA $\underline{88}$: 1760-1764.

18. Berry, A., DeVault, J.D. and Chakrabarty, A.M. (1989). J. Bacteriol. $\underline{171}$:2312-2317.

19. DeVault, J.D., Kimbara, K. and Chakrabarty, A.M. (1990). Molec. Microbiol. $\underline{4}$:737-745.

20. Kimbara, K. and Chakrabarty, A.M. (1989). Biochem. Biophys. Res. Commun. $\underline{164}$:601-608.

21. Sangodkar, U.M.X., Aldrich, T.L., Haugland, R.A., Johnson, J., Rothmel, R.K., Chapman, P.J. and Chakrabarty, A.M. (1989). Acta Biotechnol. $\underline{9}$:301-316.

22. Ghosal, D., You, -I.S., Chatterjee, D.K. and Chakrabarty, A.M. (1985). Science $\underline{228}$:135-142.

23. Frantz, B. and Chakrabarty, A.M. (1987). Proc. Natl. Acad. Sci. USA $\underline{84}$:4460-4464.

24. Aldrich, T.L. and Chakrabarty, A.M. (1988). J. Bacteriol. $\underline{170}$:1297-1304.

25. Rothmel, R.K., Aldrich, T.L., Houghton, J.E., Coco, W.M., Ornston, L.N. and Chakrabarty, A.M. (1990). J. Bacteriol. $\underline{172}$:922-931.

26. Rothmel, R.K., Haugland, R.A., Coco, W.M., Sangodkar, U.M.X. and Chakrabarty, A.M. (1989). In Recent Advances in Microbial Ecology (T. Hattori, Y. Ishida, Y. Maruyama, R.Y. Morita and A. Uchida, Eds.), Japan Scientific Societies Press, Tokyo, p. 605-610.

27. Schlomann, M., Pieper, D.H. and Knackmuss, H.-J. (1990). In: Pseudomonas: Biotrans- formations, Pathogenesis, and Evolving Biotechnology (S. Silver, A.M. Chakrabarty, B. Iglewski and S. Kaplan, Eds.), American Society for Microbiology, Washington, D.C., p.185-196.

28. Ghosal, D., You, -I.S., Chatterjee, D.K. and Chakrabarty, A.M. (1985). Proc. Natl. Acad. Sci. USA $\underline{82}$:1638-1642.

29. Ghosal, D. and You, -I.S. (1989). Gene $\underline{83}$:225-232.

30. Sangodkar, U.M.X., Chapman, P.J. and Chakrabarty, A.M. (1988). Gene $\underline{71}$:267-277.

31. Tomasek, P.H., Frantz, B. Sangodkar, U.M.X., Haugland, R.A. and Chakrabarty, A.M. (1989). Gene $\underline{76}$:227-238.

32. Haugland, R.A., Sangodkar, U.M.X. and Chakrabarty, A.M. (1990). Mol. Gen. Genet. $\underline{220}$:222-228.

33. Haugland, R.A., Sangodkar, U.M.X., Sferra, P.R. and Chakrabarty, A.M. (1991). Gene $\underline{100}$: 65-73.

BACTERIAL HEAVY METAL DETOXIFICATION AND RESISTANCE SYSTEMS

Simon Silver

University of Illinois, Chicago, IL 60680 USA

ABSTRACT

Bacterial plasmids contain genetic determinants for resistance systems for Hg^{2+} (and organomercurials), Cd^{2+}, AsO^{2-}, AsO_4^{3-}, CrO_4^{2-}, TeO_3^{2-}, Cu^{2+}, Ag^+, Co^{2+}, Pb^{2+}, and other metals of environmental concern. In some cases , there is the potential for using genetically engineered microbes for bio-remediation. Recombinant DNA analysis has been applied to mercury, cadmium, zinc, cobalt, arsenic, chromate, tellurium and copper resistance systems. The eight mercury resistance systems that have been sequenced all contain the gene for mercuric reductase, the enzyme that converts toxic Hg^{2+} ions to less toxic volatile metallic Hg^0. Four of these systems also determine the enzyme organomercurial lyase, which cuts the Hg-C bond and thus detoxifies methylmercury and phenylmercury. Two sequenced Cd^{2+} resistance determinants govern cellular efflux of Cd^{2+} assuring a low level of intracellular Cd^{2+}: not an obvious candidate for bioremediation. Cadmium accumulation by bacterial metallothionein or phytochelatin is a potentially useful process, but only preliminary reports have appeared on bacteria producing polythiol polypeptides. For arsenic resistance, a unique efflux ATPase maintains low intracellular As levels. A bacterial AsO^{2-} oxidase has been reported, with the potential of converting more toxic As(III) into less toxic As(V), but this system has not been studied in recent years. For chromate, resistance results from reduced cellular uptake. However, both soluble and membrane-bound Cr(VI) reductase bacterial activities convert more toxic Cr(VI) to less toxic Cr(III) in different bacteria.

INTRODUCTION

Bioremediation of heavy metal wastes by either bioaccumulation or by enzymatic detoxification has not been put to practical use. The potential for the former has been demonstrated by biotechnology startup companies (Brierley et al., 1986, 1989; Hutchins et al., 1986; Darnall, 1989; Darnall et al., 1989). For enzymatic detoxification, mercury-contaminated sludge has effectively been decontaminated at a laboratory scale (Hansen et al., 1984; Hansen, personal communication). Here in a Symposium on Environmental Biotechnology (in particular as related to problems faced by newly industrializing nations), I will describe the potential for these two quite different methods for bioremediation of toxic

Biotechnology and Environmental Science: Molecular Approaches
Edited by S. Mongkolsuk et al., Plenum Press, New York, 1992

109

mineral wastes: firstly bioaccumulation and then the conversion of toxic heavy metals by oxidation or reduction from a more toxic form to a less toxic (or readily removed from the environment) form. Three such bioconversions are reduction of Hg(II) and Hg(I) ions to metallic Hg(0), the oxidation of more toxic arsenite [As(III)] to arsenate [As(V)]; and the reduction of more toxic chromate [Cr(VI)] to less toxic and more readily removed chromium ions [Cr(III)] . The molecular biology of bacterial resistances to mercury, arsenic and chromate has recently been developed in considerable detail, and that information will be summarized. For the mercury system, the resistance mechanism is the detoxification mechanism. For currently known examples of arsenic and chromate resistances, microbial plasmid-determined heavy metal resistances do not involve biotransformations, but rather membrane transport mechanisms that assure lower intracellular concentrations for the toxicant than is found in the external environment. (Related reviews of this area will appear in the published proceedings two conferences, the U.S.-Israel Conference on Advances in Applied Biotechnology, Haifa, Israel, and the Eight International Biodeterioration Symposium, Windsor, Canada, both held also during the Summer of 1990).

CADMIUM RESISTANCE SYSTEMS OF BACTERIAL PLASMIDS

There are three well-defined systems for bacterial cadmium resistance (plus several others apparently involving cadmium-binding components; see next section). These are the cadA and cadB determinants of Staphylococcal plasmids (Novick et al., 1979; Silver and Laddaga, 1990) and the czc (for cadmium, zinc and cobalt) resistance system of the Gram negative soil microbe Alcaligenes eutrophus (D. Nies et al., 1987, 1989). The cadA and cadB systems also confer resistance to zinc, although still other Staphylococcal cadmium resistance systems are specific for cadmium alone (Witte et al., 1986).

The cadA cadmium (and zinc) resistance system works by means of reduced cellular accumulation of cadmium (Weiss et al., 1987). Cd^{2+} uptake occurs in Gram positive bacteria via the membrane manganese transport system, and since all cells need manganese for growth and enzyme activity, all cells have a highly specific Mn^{2+} carrier.Cd^{2+} is an undesired alternative substrate for this membrane active transport system. Cadmium resistant chromo-somal mutants occur that have lost the specificity for Cd^{2+} for the chromosomal system (Laddaga and Silver, 1985). The lesser cadmium accumulation by plasmid-containing resistant cells is not due to a direct block on uptake but rather due to accelerated efflux of Cd^{2+} (Tynecka et al., 1981). The system responsible for cadmium resistance and efflux was cloned and sequenced (Nucifora et al., 1989b). It contains two open reading frames. The polypeptide product of the longer gene (cadA) has the characteristics of a P-class ATPase (Nucifora et al., 1989b; Silver et al., 1989). The second open reading frame (called cadC since cadB was saved for the second cadmium resistance system on the same plasmid; Novick et al., 1979) was recently shown to be essential for full cadmium resistance (Yoon and Silver, 1991). From its amino acid sequence, CadC appears to be a soluble protein, perhaps loosely associated with the CadA membrane ATPase, as is often the case with multicomponent P-class ATPase.

The current model for the CadA membrane ATPase (Fig. 1) depends heavily upon what is known about the better studied members of this class, the Ca^{2+} ATPase of animal sarcoplasmic reticulum and the Na^+/K^+ ATPase of animal membranes (Silver et al., 1989). The P-class ATPases are all cation transporting ATPases, which are phosphorylated during the reaction cycle at a conserved aspartate residue (Asp415 in the Cd^{2+} ATPase). The cation substrate can differ (Ca^{2+}, Mg^{2+}, Na^+, K^+, H^+, or Cd^{2+}) and the direction of transport (uptake or efflux) can

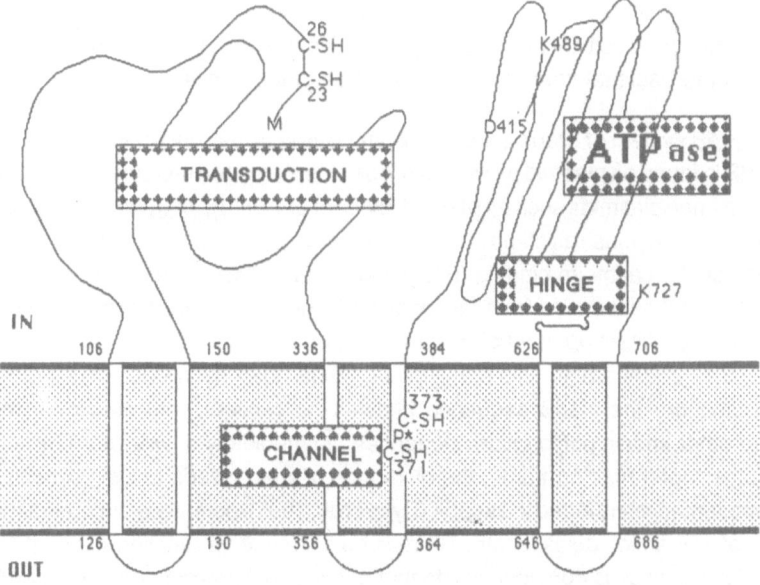

FIG. 1 Model of the cadmium ATPase of the Cd²⁺ Zn²⁺ resistance system of Gram positive bacteria (from Silver et al., 1989, with permission).

differ as well. The CadA ATPase is thought to contain several functional domains, starting with a region including Cys23 and Cys26 that functions in cadmium binding. This region is homologous in sequence to the mercury periplasmic protein and the N-terminus of mercuric reductase (see below), both of which are thought to play similar roles in recognizing and binding Hg^{2+} ions (Silver and Misra, 1988). Then follows a membrane "hairpin" region, where the polypeptide passes back and forth across the cell membrane, so that very few residues are extracellular. Next comes a region referred to as a "transduction" domain (Fig. 1), because it is thought to guide the cation substrate to the membrane channel. This region is seen as a "stalk" in electron micrographs of the well-studied Ca^{2+} ATPase. It is also referred to as a "phosphatase" domain, as it may be involved in dephosphorylating phospho-aspartate$_{415}$. The transduction domain is about 190 amino acids in length. Continuing along the sequence, the next membrane hairpin probably is involved in the membrane transport channel itself (Fig. 1). It contains a conserved proline residue (Pro372), which falls between the only additional cysteine residues (Cys371 and Cys373) in this long, unusually cysteine-poor polypeptide. Next comes the 250 amino acid ATPase domain, which includes the phosphorylation site, Asp415, in a conserved stretch of 7 amino acids that are the same in all of these enzymes, regardless of substrate and direction, from bacteria to plant and animal. The Lys489 residue is homologous to one protected from chemical attack by ATP binding in the calcium enzyme. The "hinge" region (at the end of the ATPase domain and including the next hydrophobic, apparently membrane-spanning a-helical region) is the most conserved region in the P-class enzymes (Silver et al., 1989). After the third membrane-spanning hairpin, the cadmium ATPase ends. Some members of this enzyme class in eukaryotes continue for 200-300 additional amino acids, with one or two additional membrane hairpins. This physical model from computer analysis and homology to related animal enzymes provides a detailed functional model as well. In addition to the cadA and cadC genes, which make up the CadA ATPase determinant, the starting point for mRNA synthesis was determined (Yoon et al., 1991). The CadA resistance system is produced only when induced by Cad²⁺ or other divalent cations. Gene fusion experiments have led to the hypothesis that there must be a third gene, cadR, to regulate gene activity (Yoon et al., 1991). However, the location of this gene on the plasmid or on the bacterial chromosome has not be pinned down.

The cadB system of Staphylococcal plasmids is less well understood than is the cadA system. cadB is found on plasmid pII147 but not on closely related mercury-cadmium-arsenic-pencillin-resistance plasmids such as pI258 (Novick et al., 1979). cadB maps as a completely separate non-homologous resistance determinant from cadA. Its mechanism of resistance may involve cellular cadmium binding (Perry and Silver, 1982). The cadB determinant was recently cloned and sequenced from the small cadmium resistance plasmid pOX6 (Dyke et al., 1991). Two significant open reading frames were required for Cd^{2+} resistance. The cadB gene determines a 204 amino acid polypeptide. Other than hydrophobicity, indicative of a membrane location, the CadB sequence showed no recognizable functional motifs nor homology to other known polypeptide sequences. Immediately downstream from the cadB gene is another open reading frame, the disruption of which leads to Cd^{2+} sensitivity. This reading frame, tentatively referred to as cadX (Dyke et al., 1991), potentially encodes a hydrophilic polypeptide of 116 amino acids length. There is no evidence for functional role of CadX, but its amino acid sequence is 40% identical to that of the CadC polypeptide of the cadA system (Yoon and Silver, 1991) and about 30% identical to the ArsR regulatory polypeptides of both Gram negative (San Francisco et al., 1990) and Gram positive Staphylococcal (Rosenstein and Götz, 1991) origin. There is a problem with interpreting these sequence homologies in the absence of more direct evidence for protein function. The CadC polypeptide of the cadA system cannot be required for regulation, since cadC-bla fusions are still inducible by cadmium (Yoon and Silver, 1991); ArsR has been shown to be needed for regulation of but not function of the arsenic resistance operons (Tisa and Rosen, 1990; F. Götz, personal communication). Therefore the function of CadX cannot currently be assigned.

The czc (cadmium, zinc and cobalt) resistance system of Alcaligenes plasmid pMOL30 was cloned (Nies et al., 1987) and shown to determine the efflux of all three toxic substrates (Nies and Silver, 1989). The system functions inducibly. The czc DNA sequence has been completed (D. Nies et al., 1989) and shows four open reading frames (czcC, czcB, czcA, and czcD in order) that correspond to proteins seen on radioactive gels. There is no recognizable ATP binding motif on any of the sequences. CzcA, at 1064 amino acids, is a long very hydrophobic membrane protein that is essential for the functioning of the czc system. The czcD gene was not essential for resistances or cation efflux with the cloned system (D. Nies et al., 1989). Internal deletion mutants in czcB retained partial cobalt resistance, but had lost cadmium and zinc resistances; and internal mutations in czcC caused loss of cadmium resistance (and efflux) but retention of zinc resistance and efflux. Thus the czcB and czcC genes appear to affect the cation substrate range of a efflux system built around the central CzcA membrane protein.

COPPER RESISTANCE SYSTEMS ON BACTERIAL PLASMIDS

Copper resistance in bacteria has been observed periodically (Trevors, 1987) and quite distinct plasmid copper resistance determinants are currently under careful analysis. However, the basic mechanism(s) of these plasmid resistance systems in Pseudomonas and E. coli are not understood.

The copper resistance determinant of plant-pathogenic Pseudomonas was cloned and shown to be quite widespread where copper-salts were sprayed for plant pest control

(Cooksey, 1987). The DNA sequence of the copper-resistance determinant contained four significant open reading frames (Mellano and Cooksey, 1988). The first two reading frames were required for resistance, but the second two were needed only for full copper resistance. An octapeptide (Asp-His-Ser-Gln/or Lys-Met-Gln-Gly-Met) sequence was found 5 times in the product of the second open reading frame, and related less conserved octapeptide sequences appeared four times in the first open reading frame. Basically, the DNA analysis of this determinant of copper resistance has moved faster than the physiological and biochemical understanding of the mechanism involved. Another plasmid determinant of copper resistance from a related plant-pathogenic bacterial species also occured wide-spread but did not show detectable DNA hybridization with the Pseudomonas system (Bender et al., 1990).

The E. coli plasmid determinant of copper resistance appeared in the effluent from piglets fed copper salts as a growth stimulant (Tetaz and Luke, 1983). The system has been cloned and governs an inducible resistance (Rouch et al., 1985) that involves copper binding and copper transport (Rouch et al., 1989a). The net result is decreased net copper accumulation, but like the chromate resistance system described below, the data are currently unclear as to whether a block on uptake or accelerated efflux is responsible. The copper resistance determinant was cloned and four genes were identified; they seem unrelated to the reading frames of the Pseudomonas plasmid. The first, pcoC, gene determines a 26 kDa intracellular copper-binding protein (Fig. 2). The next gene pcoR determines a trans-acting regulatory factor, and the remaining two genes pcoB and pcoA have products that are components of a membrane copper efflux system (Rouch et al., 1989a and b; Lee et al., 1990) (Fig. 2). The copper resistance system differs from many others described in this review, since (i)copper is both a required nutrient (at low levels) and a toxic pollutant (when present in excess), and (ii) both plasmid and chromosomal gene products interact in copper transport, intracellular binding and efflux (Fig. 2; Rouch et al., 1989b; Lee et al., 1990). Although the model in Fig. 2 of two parallel chromosomally-determined copper uptake systems (CutA and CutB), intracellular copper processing proteins determined by both plasmid (pcoC) and chromosomal (cutE and cutF) genes, and a copper efflux system with both chromosomal (cutC and cutD) and plasmid gene determination is quite precise, it is very tentative because the

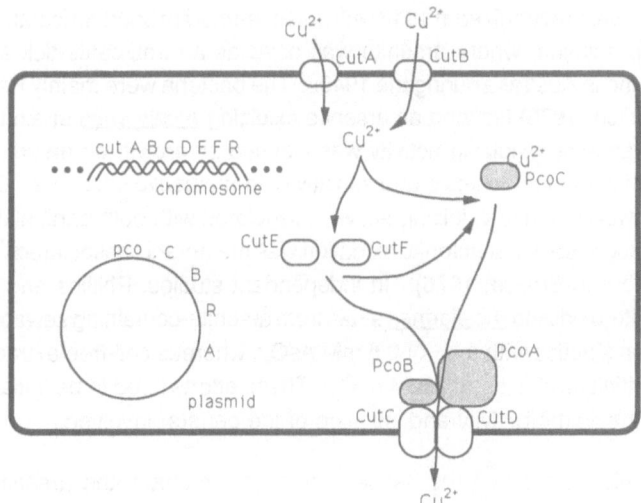

FIG. 2   The copper uptake and efflux processes in Escherichia coli (modified from Rouch et al., 1989a and b, Lee et al., 1990, and personal communication).

basic data upon which it is based have not appeared. The cloned E. coli copper resistance system is currently being sequenced (D. Rouch and N. L. Brown, personal communication), so we can anticipate better understanding soon.

Periodically, reports appear of the production of excess hydrogen sulfide and associated extracellular precipitation of cation sulfide complexes, as a basis for bacterial resistance and detoxification of such materials as Hg^{2+}, Cu^{2+} and Cd^{2+}. Copper sulfide precipitation was reported as the mechanism for plasmid-determined copper resistance in a Mycobacterium scrofulaceum isolate (Erardi et al., 1987) for which the same plasmid determines mercury resistance by volatization (i.e. reduction) (Meissner and Falkinham, 1984). A black copper-sulfide precipitate formed on resistant, plasmid-containing cells but not on sensitive, plasmid-free cells (Erardi et al. 1987). Whether the cells are precipitating (as HgS) and volatilizing (as Hg^0) mercury simultaneously is not clear, but removal of mercury by either means should enhance bacterial growth. Aiking et al. (1985) showed that a Klebsiella aerogenes isolate precipitated (and therefore detoxified) mercury, cadmium and lead salts, by a process dependent upon added sulfate, or rather the sulfide made by the cells from sulfate . Perhaps the surprise is that this mechanism of cation-sulfide precipitation has been reported so seldom as the mechanism of toxic cation resistance.

ARSENITE OXIDATION AND RESISTANCE

I know of no microbial system for reduction of As(III) or As(V) to As(0). If such a system exists, analogous to the microbial mechanism for reduction of sulfate to elemental sulfur, then it might function as a useful system for bioremediation. There is a much understanding of the molecular biology and biochemistry of plasmid determined As(III) and AS(V) resistances and less on the oxidation of more toxic arsenite [As(III)] to less toxic arsenate [As(V)]. Unfortunately, plasmid-determined resistance does not involve redox chemistry (see below).

In the dozen years since we reviewed the literature on microbial arsenite oxidation (Summers and Silver, 1987), there have been no additional studies on the biochemistry or genetics of this process. Such studies are essential, if the process is to be used. Arsenite oxidizing bacteria were recognized in 1918 with arsenite-tolerant bacteria isolated from a South African cattle dip solution, where arsenite was used as an anti-cattle-tick agent. Similar bacteria were found in Australia during the 1940s. The bacteria were mainly Pseudomonads. Osborne and Ehrlich (1976) isolated an arsenite oxidizing Alcaligenes strain from soil. The synthesis of the arsenite oxidizing activity was induced by growth on arsenite. Whole cell oxidation showed saturation kinetics and sensitivity to metabolic inhibitors. Crude cell-free preparations showed arsenite oxidizing activity associated with both particulate and soluble fractions. It was suggested that arsenite oxidation was membrane-associated via cytochrome oxidase (Osborne and Erhlich, 1976). In independent studies, Phillips and Taylor (1976) isolated an arsenite-oxidizing Alcaligenes strain from arsenite-containing sewage. Whole cells showed saturation kinetics with a K_m of 0.5 mM AsO_2^-, whereas cell-free extracts showed an 100-fold higher affinity with a K_m of 2 mM AsO_2^-. These efforts need to be initiated again with purification of the enzyme activity and isolation of the gene(s) involved.

The plasmid determined resistance system for arsenate and arsenite (which also confers resistance to antimonate) has been studied both with E. coli and with S. aureus. Earlier results were summarized by Silver and Misra (1988). Resistance to arsenite did not involve oxidation to arsenate (Silver et al., 1981). There was no evidence of extracellular detoxification or chelation of arsenite or arsenate. Resistant cells accumulated less arsenate than sensitive cells--and that alone accounted for the resistance (Silver et al., 1981). The reduced accumulation was due to rapid efflux of accumulated oxyanions from the cells by an inducibly-synthesized plasmid-governed system (Silver and Keach, 1982; Mobley and Rosen, 1982).

The molecular genetics of the arsenate/arsenite/antimony resistance determinant of plasmid R773 (from Gram negative bacteria) has progressed recently with purification of the three protein components, cloning and sequencing of the genes, and mutational analysis of gene functions and component interactions. This work was recently summarized by Tisa and Rosen (1990). The three genes that together determine the arsenic resistance efflux ATPase were cloned and sequenced by Chen et al. (1986). The additional upstream regulatory gene arsR was identified by San Francisco et al. (1990). Transcription into mRNA starts at the left at a promoter identified by Owolabi and Rosen (1990) and continues for 4.4 kb. The first gene, arsR, determines the 117 amino acid regulatory protein, which is 30% identical in sequence to the comparable gene product from a Gram positive Staphylococcal plasmid (R. Rosenstein and F. Götz, 1991) (Fig. 3). After the arsR gene is a 1 kb region of DNA lacking open reading frames (San Francisco et al., 1990), but with a potential regulatory inverted repeat sequence. Then follow the three genes for component of the ATPase, arsA (for the ATPase subunit itself), arsB (for an integral membrane protein, presumedly containing the oxyanion transmembrane channel) and arsC (for a small soluble polypeptide needed for arsenate--but not for arsenite--resistance) (Chen et al., 1986; Tisa and Rosen, 1990). There are inverted repeated sequences as candidates for regulatory control between the arsA and arsB genes and after the arsC gene as well. The ArsA amino acid sequence is unusual in that it appears to be a tandem fused dimer, resulting from gene duplication and fusion (Chen, et al., 1986). Both halfs have ATP-binding motif sequences. Mutational analysis showed that the N-terminal motif $G_{15}KGGVGKTS_{23}$ is required for efflux activity and resistance (Karkaria and Rosen, 1990). It is not known whether the second ATP-binding motif is essential and whether both play enzymatic roles, or one region is enzymatic and the other regulatory. The ArsA ATPase subunit has been purified and characterized physically (Hsu and Rosen, 1989). It is an arsenite (and antimonate)-stimulated ATPase (Rosen et al., 1988). Conformational changes upon addition of ATP or oxyanions was indicated by protection from proteolysis (Hsu and Rosen, 1989). The next gene product, ArsB is a membrane protein (San Francisco et al., 1989) that is made in smaller amounts than ArsA (Tisa and Rosen, 1990), apparently because of mRNA processing leading to a shorter mRNA missing the arsB segment (Owolabi and Rosen, 1990). The ArsB protein anchors the otherwise soluble ArsA ATPase subunit to the cell membrane (Tisa and Rosen, 1989). The complex of ArsA and ArsB suffices to confer resistance to (and

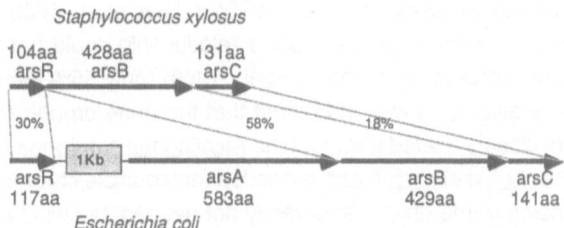

FIG. 3 Comparison of arsenic resistance determinants of Staphylococcus and E.coli (modified from Silver and Laddaga, 1990, with permission).

presumedly efflux of) arsenite and antimonate, but the third gene arsC is needed for arsenate resistance (and efflux) (Chen et al., 1986). The function of ArsC hypothesized to provide specificity on the transport system in a manner analogous to periplasmic substrate-binding proteins, but on the inside of the cell for a transport system functioning in efflux. The ArsC amino acid sequences from plasmid R773 and the Staphylococcal system (below) are weakly homologous (about 15% amino acid identities) to the predicted translation product of an open reading frame involved in nitrogen fixation in Azotobacter that is now in the gene product libraries.

The arsenite/arsenate/antimonate resistance determinant of S. aureus plasmids appears quite similar in function to that of Gram negative bacteria (Silver et al., 1981). Energy-dependent arsenic efflux is the mechanism of resistance for both Gram positive and Gram negative bacteria (Silver and Keach, 1982). Recently, the Staphylococcal arsenic resistance operon has been cloned and sequenced (Rosenstein and Götz, 1991) (Fig. 3). The results are quite surprising. Although the arsR genes with which both systems start are homologous so that 30% of the amino acid translation products are identical (Fig. 3), Staphylococcal sequence is missing the arsA gene. There is no room within the sequence for an additional gene and the length of DNA/DNA homology between the sequenced Staphylococcal system and the more throughly studied one of plasmid pI258 (Götz et al., 1983; Silver et al., 1981) has no room for the arsA gene. We are left with unsatisfactory alternative hypotheses that (a) arsA in Staphylococcus lies on the chromosome rather than on the plasmid, or (b) that ArsB alone functions as an oxyanion efflux system in Staphylococcus, without the need for the ATPase subunit ArsA. The ArsB amino acid sequences are 58% identical between the Staphylococcal and enteric versions of this gene product (Fig. 3), and the regions of homology are spread uniformly from N- to C-end. The Staphylococcal sequence has an arsC gene as well, whose amino acid translation product is 18% identical to that from plasmid R773 (Fig. 3), significantly but weakly homologous. The loss of arsC from the Staphylococcal system also results in loss of arsenate resistance, but not arsenite resistance (Rosenstein and Götz, 1991).

CHROMATE REDUCTION AND RESISTANCE

For chromate (as for arsenite), there is both microbial detoxification (reduction of chromate to Cr^{3+}) and plasmid-determined chromate resistance: but the two processes are unrelated. Chromate resistance results from lowered accumulation by the cells (Ohtake et al., 1987). It is not yet clear whether this results from a direct block on uptake (via the chromosomally-determined sulfate transport system) or accelerated efflux following uptake.

Two plasmid-determined chromate resistance systems were recently cloned and sequenced in our laboratory (Cervantes et al., 1990; A. Nies et al., 1989, 1990). Since the DNA sequences were sufficiently different that a relationship could not be recognized by Southern DNA/DNA hybridization (Cervantes et al., 1990) (only 55% identical bases, when the sequences were available), it was of interest that the gene organization and predicted protein products were closely related (Fig. 4). The Alcaligenes eutrophus system (as cloned) consists of two genes chrA and chrB that are needed for inducible chromate resistance plus third partial open reading frame (that is apparently not needed, but which is complete in the second, Pseudomonas aeruginosa chromate resistance system). The long 401 or 416 amino acid ChrA protein is found in the cell membrane (Cervantes et al., 1990; A. Nies et al., 1990). It may be the only structural component of the membrane transport resistance determinant.

116

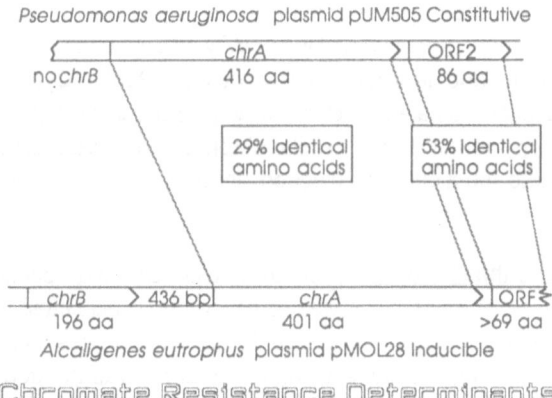

Pseudomonas aeruginosa plasmid pUM505 Constitutive

Alcaligenes eutrophus plasmid pMOL28 Inducible

Chromate Resistance Determinants

FIG. 4 Comparison of plasmid chromate resistance determinants from Pseudomonas and Alcaligenes (from Cervantes and Silver, 1990, with permission).

Other than hydrophobicity, the amino acids sequences give no suggestion for functional regions, and in particular the sequences lack discernable ATP-binding motifs. The upstream chrB gene of the Alcaligenes system is likely to be the regulatory gene, since it is present in the cloned Alcaligenes system (inducible) (A. Nies et al., 1990) but absent in the cloned Pseudomonas system (constitutive) (Cervantes et al., 1990). A chrA-lacZ gene fusion responds specifically to chromate induction, but not to other tested oxyanions (D. Nies, personal communication). Whether gene regulation occurs upstream of the chrB gene and/or within the 436 base pairs between chrB and chrA needs to be determined. The shorter 86 codon open reading frame in the Pseudomonas clone and sequence is apparently not required for chromate resistance, because it was disrupted during cloning of the Alcaligenes system. Yet, 53% of the first 69 amino acids in the predicted gene products are identical (fig. 4). Clearly, work has just started on this new system and there is much to learn.

The P. fluorescens strain that provided the first chromate-resistance plasmid also has the capacity to reduce chromate to Cr^{3+} (Bopp and Ehrlich, 1988), but the two phenomena are not related: a strain cured of plasmid retains chromate reductase activity but has lost chromate resistance (Bopp and Ehrich, 1988; Ishibashi et al., 1990). Nevertheless, the potential for use of this microbe to remove toxic chromate from water was patented (Bopp, 1984).

Progress toward development of a useful chromate reductase activity recently has been achieved by Ohtake and co-workers with an anaerobic membrane respiratory-chain system from a newly-isolated Enterobacter cloacae. This chromate-resistant E. cloacae strain was found in activated sludge (P.C. Wang et al., 1989). Although, it is resistant to chromate under both aerobic and anaerobic conditions, chromate reduction occurs only anaerobically (P.C. Wang et al., 1989). Resistance to and the ability to reduce chromate were lost after anaerobic growth on nitrate, suggesting that the nature of the anaerobic respiratory pathway is crucial for chromate reduction. The reduced chromate formed insoluble chromium hydroxide. Chromate reductase activity was inhibited by oxygen (P.C. Wang et al., 1989; Komori et al., 1989, 1990a). Initial levels of 1 or 2 mM CrO_4^{2-} were rapidly detoxified, although higher concentrations inhibited cellular activity. A variety of carbon sources functioned as electron donors for chromate reduction and the activity was inhibited by other oxyanions such as molybdate, vanadate and tellurate (Komori et al., 1989). Cyanide and membrane uncouplers also inhibited chromate reductase activity. Subcellular right-side-out membrane vesicles from E. cloacae contained the cellular chromate reductase activity and ascorbate-phenazine methosulfate functioned as electron donor with membrane preparations (P.C. Wang et al., 1990).

Attempts to use the E. cloacae strain to detoxify chromate-containing effluents are being made. Up to 90% of 4mM CrO_4^{2-} was removed and precipitated within dialysis bags containing intact bacteria (Komori et al., 1990a, b), but the process took several days. For reduction of waste chromate, a small supplement of broth was needed to provide energy (Ohtake et al., 1990). It was also sometimes necessary to dilute the chromate-containing waste prior to exposure to the bacteria, presumedly to lower the levels of other toxic materials present. These efforts are preliminary and further work using bacteria within dialysis bags or cells immobilized on columns may lead to the development of a useful chromate detoxification technology.

Ishibashi et al. (1990) provided the newest (and still more preliminary) description of bacterial soluble chromate reductase activity (Fig. 5) that functions aerobically. The K_m for this system was about 40 mM CrO_4^{2-}, and the activity was highly specific for chromate. Sulfate and nitrate were without effect, although mercury and silver cations noncompetitively inhibited chromate reductase activity. This system may be identical to that reported by Horitsu et al. (1987) for a different Pseudomonad isolate. There have been several additional reports of chromate reduction as a possible means of bioremediation (e.g. Romanenco and Kkoren'kov, 1977; Lebedeva and Lyalikova, 1979; Horitsu and Kato, 1980; Shimada and Matsushima, 1983; Kvasnikov et al., 1985; Gvozdyak et al., 1986) but these have not been developed further.

Although several microbial activities with mercury and mechanisms for bacterial resistances to mercury have been reported periodically (reviewed by Belliveau and Trevors,

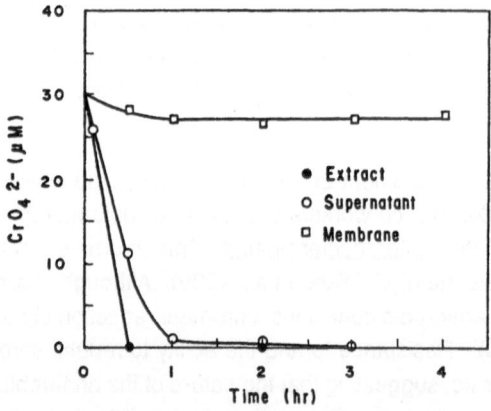

FIG. 5 Chromate reduction by cell fractions from Pseudomonas putida (from Ishibashi et al., 1990; with permission).

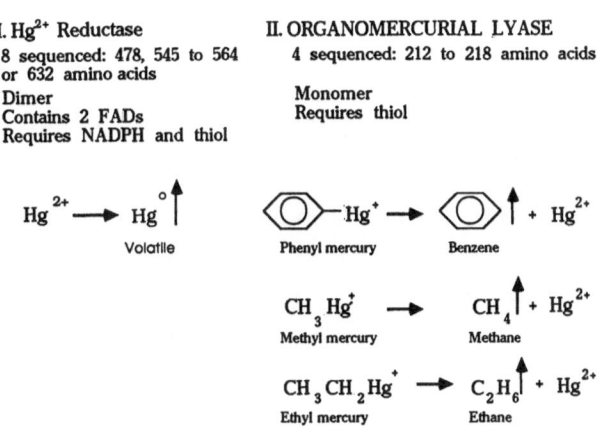

I. Hg^{2+} Reductase
8 sequenced: 478, 545 to 564 or 632 amino acids
Dimer
Contains 2 FADs
Requires NADPH and thiol

II. ORGANOMERCURIAL LYASE
4 sequenced: 212 to 218 amino acids
Monomer
Requires thiol

FIG. 6 The enzymes for mercury and organomercurial degradation.

MERCURY REDUCTION AND VOLATILIZATION

1989), our laboratory has found only the single process for mercury resistance (reduction from Hg^{2+} to less toxic and volatile Hg^0) in studies of more than 200 mercury-resistance strains in a wide variety of Gram positive and Gram negative bacteria. Thus, we focus on this mechanism.

There has been considerable progress on the molecular genetics of the mercurial detoxification systems (next below), but efforts in using these systems for bioremediation of toxic mercurial wastes have been slow. Additional work needs to be done here. With mercurial-polluted soil, the addition of mercury-volatilizing P. aeruginosa cells (with or without added nutrients) had little effect on mercury elimination, presumedly because the soil-bound mercury is well fixed on particles and inaccessible to the bacteria (Silver, unpublished data). Addition of soluble thiol compounds may potentiate bioremediation of mercury-polluted soil.

With well-stirred aqueous systems, there is the essentially complete and continuous elimination of inorganic mercury waste by the bacterial system (Hansen et al., 1984). In more recent efforts with bioreactors, Hansen (personal communication) has been able to to achieve long term (up to 2 weeks) continuous removal of mercury from 75 mg/l (0.37 mM) down to 120 mg/l (0.6 mM; more than 99% removal). The reduced mercury could be recovered either as a sludge containing essentially 50% Hg^0 by weight or was volatilized from the system, depending upon aeration. Volatilized Hg^0 could be recovered on an activated charcoal column, if desired.

By far the best understood heavy metal resistance system is that for mercuric ions and organomercurial compounds. The mechanism of resistance in bacteria is enzymatic detoxification. More toxic Hg^{2+} is converted to less toxic Hg^0, which is volatilized away from the contaminated site (and the bacteria). Organomercurials such as phenylmercury and methylmercury are first cleaved enzymatically to inorganic Hg^{2+} plus an organic compound (benzene and methane respectively for the two examples). The enzymes responsible for these conversions are mercuric reductase and organomercurial lyase (Fig. 6).

Eight mercury resistance systems have been cloned and their DNA sequences

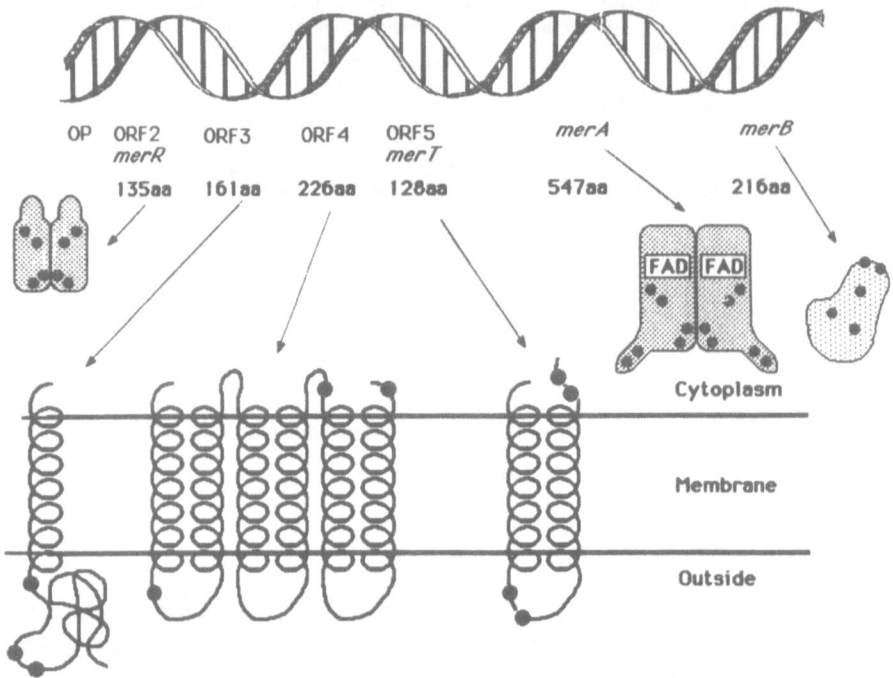

FIG. 7 The mercuric/organomercurial resistance system of Staphylococcus (from Silver and Laddaga, 1990, with permission).

determined. In each case, the system consists of a number of genes (approximately 6), whose products carry out a carefully regulated process of mercury uptake by the cell, delivery of the toxic Hg^{2+} ions to the intracellular enzyme mercuric reductase, with the subsequent volatilization of the product, gaseous metallic Hg^0.

Fig. 7 shows the organization of this genetic system in the Gram positive bacterium Staphylococcus aureus. While the operon organization is somewhat different in other Gram positive and Gram negative bacteria, the overall process remains the same. The mer operon begins with a regulatory gene, merR, the product of which is a 135 amino acid polypeptide that functions in the cell as a dimer (O'Halloran et al., 1989; O'Halloran, 1989) and binds to the upstream operator/promoter segment (itself less than 75 base pairs long). In the absence of Hg^{2+} ions, the MerR protein represses mRNA synthesis of the next (structural) genes. In the Gram negative systems, MerR also auto-regulates its own synthesis from a separate promoter in the same region (Lund and Brown, 1989). When Hg^{2+} ions are added, the MerR protein functions positively to stimulate mRNA synthesis more than 100x fold, 'turning on" the mercuric resistance system. The Bacillus MerR dimer has been shown to form a tri-cysteine complex with a single Hg^{2+} ion bound by Cys_{79} on one subunit and Cys_{114} and Cys_{123} on the other subunit (Helmann et al., 1990). With MerR responding to organomercurials as well as to inorganic Hg^{2+}, a deletion of the C-terminal 10 amino acids eliminated the in vivo response to organomercurials but not to Hg^{2+} (Nucifora et al., 1989a). Hg^{2+} binding is exceedingly tight and specific for Hg^{2+} in vivo, although Cd^{2+}, Zn^{2+} and other divalent cations work weakly in vitro (Ralston and O'Halloran, 1990; Helmann and Walsh, 1990).

Then follows a series of 4 to 6 genes (5 in <u>Staphylococcus</u>; Fig. 7), whose products carry out the transport and detoxification of mercury. The next three <u>mer</u> genes are involved in mercury transport, but how these gene products interact is unclear (Silver and Laddaga, 1990). In Gram negative bacteria, the homolog to ORF5 is <u>merT</u>, the product of which is an integral membrane protein that functions as a Hg^{2+} transport system (Kusano et al.,1990), bringing mercury from outside of the cell to inside where the mercuric reductase enzyme resides. Since intracellular Hg^{2+} would be toxic, it is proposed that the mercury bound to MerT is directly transferred to the mercuric reductase enzyme, which may be loosely associated with the cellular membrane. The N-terminal portion of mercuric reductase shares a sequence homology to the MerP periplasmic polypeptide (of the <u>mer</u> operons of Gram negative bacteria) and to the beginning of the <u>cadA</u> Cd^{2+} ATPase (of Gram positive bacteria), thus leading to the suggestion that it functions intracellularly as part of the mercury "bucket bridge" from the outside, across the membrane, eventually to the active site of the mercuric reductase enzyme.

The Hg^{2+} ion is then transferred from the N-terminal cysteine pair of the mercuric reductase enzyme to a highly conserved cysteine pair at the C-terminus. These cysteines are required for enzymatic activity (Moore and Walsh, 1989). It is thought that the Hg^{2+} may be tetragonally coordinated (an unusual structure for mercury) between the carboxyl-terminal cysteine pair located at the active site of the enzyme (Silver and Misra, 1988; Walsh et al., 1988; Distefano et al., 1989). It was recently shown that the mercury bridge is shared between the subunits of the dimer, since mutant heterodimers (missing one pair of cysteines --Cys_{135}, Cys_{140}-- on one subunit and the other pair of cysteines --Cys_{558} Cys_{559}-- on the second subunit) are functional both in vivo and in vitro (Distefano et al., 1990), whereas mutant homodimers do not work.

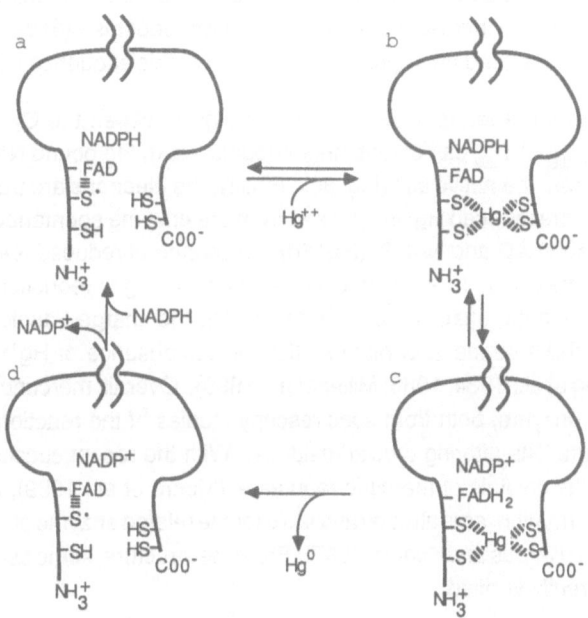

FIG. 8 Reaction mechanism of mercuric reductase (adapted from Miller et al., 1989 and Distefano et al., 1990).

The active site cysteine pair (Cys_{135} and Cys_{140}) occurs in a stretch of about 15 amino acids, which is highly conserved in all 8 currently-known versions of mercuric reductase (Silver and Misra, 1988) and in the closely related oxido-reductase enzymes glutathione reductase, trypanothione reductase and lipoamide dehydrogenase. The carboxyl-terminal cysteine pair is found in all 8 mercuric reductase sequences, but not in glutathione reductase or lipoamide dehydrogenase. The carboxyl end of glutathione reductase is, however, involved in substrate binding and is found close to the active site in the structure derived from X-ray crystallographic analysis (Thieme et al., 1981; Karpus and Schulz, 1987).

Hg^{2+} is reduced (Figs. 6-8) by electrons which have been transferred from NADPH to the enzyme-bound FAD and then to the Hg^{2+} ion (Fig. 8). The Hg^0 is released from the enzyme, diffuses rapidly from the cell (apparently without a protein carrier). In a rapidly stirred system, Hg^0 volatilizes into the atmosphere. In the absence of aeration (such as in a still ecosystem), the metallic Hg^0 remains bound to cellular lipids, hydrocarbons or sulfur-containing materials. In an unstirred anaerobic laboratory culture, the mercury appears as a gray sludge (Hansen, personal communication), associated with the cells.

The reaction mechanism for mercuric reductase has been studied in detail at the enzymological level, most recently in the laboratories of Linskog, Massey, Walsh and Williams. The activated EH_2NADPH form of the enzyme has a charge transfer complex between Cys_{140} and the bound FAD (Fig. 8a). This complex is resolvable by spectroscopic measurements (Sahlman et al, 1984, 1986; Sandstrom and Lindskog, 1987; Miller et al., 1989). The carboxyl-terminal cysteine pair Cys_{558} Cys_{559} (contributing to the active site from the other subunit; Distefano et al., 1990) is reduced. Hg^{2+} is added to the enzyme (either from soluble thiols, such as mercaptoethanol in cell-free preparations, or perhaps from the N-terminal cysteine pair Cys_{10} Cys_{13} in vivo; Brown, 1985; Silver and Misra, 1988). In vivo and in vitro, the N-terminal cysteines can be removed by mutation or with trypsin, without affecting mercury resistance or cell-free enzyme properties (Moore and Walsh, 1989). Yet, the N-terminal cysteine pair is conserved in 7 of the 8 sequenced versions of mercuric reductase (Silver and Misra, 1988) and it is difficult to understand the repeated occurrence of this sequence if it lacks a function.

Fig. 8b shows the tetragonally-coordinated Hg^{2+} between the Cys_{135} Cys_{140} on one subunit and the Cys_{558} Cys_{559} pair on the other subunit. Next, the bound NADPH is oxidized, reducing the FAD near the active site (Fig. 8c). Finally, the electrons are transferred from the $FADH_2$ to mercury; the reduced Hg^0 is released from the enzyme spontaneously, forming the enzyme with oxidized FAD and $NADP^+$ (Fig. 8d). Exchange of reduced NADPH for oxidized $NADP^+$ completes the cycle. In addition to the reaction leading to reduction of mercury ions, a process of activating the enzyme from a form lacking the charge transfer complex to one with the complex and a series of partial reactions in the absence of Hg^{2+} have been characterized (Sahlman et al., 1984, 1986; Miller et al., 1989). Overall, mercuric reductase is now a well understood enzyme, both from spectroscopy studies of the reaction mechanism and with a range of mutations altering crucial residues. With the recent success in obtaining x-ray diffraction quality crystals of mercuric reductase (Moore et al., 1989), one can hope for direct structural information, as is already available for the related enzyme glutatione reductase (Thieme et al., 1981; Karpus and Schulz, 1987). From the structure, some aspects of the model in Fig. 8 will be directly verified.

As is the case with the merA gene determining the enzyme mercuric reductase, the merB gene determining the enzyme organomercurial lyase has been identified and sequenced in plasmid pDU1358 (Gram negatives) and in the 3 Gram positive sources of the mercurial resistance operon. The amino acid sequence alignment of these polypeptides (Silver and Misra, 1988) show a family of enzymes 212 to 218 amino acids in length. The amino acid sequences are 41% to 73% identical, when compared 2 at a time. Highly conserved sequences are found throughout the polypeptide, but mostly in the central half, especially between positions 75 and 170. The lyase enzyme was purified and shown to occur as a monomer (Begley et al.,1986a). Begley et al. (1986b) and Walts and Walsh (1988) worked on the reaction mechanism of organomercurial lyase (Fig. 9), especially varying the specific substrate. A crucial cysteine (probably Cys_{96}, Cys_{117} or Cys_{159} in the E. coli sequence) initially binds the organomercurial substrate (Fig. 9b). A concerted S_E2 mechanism of carbon-mercury bond breakage and carbon-hydrogen bond formation is proposed (Begley et al., 1986b; Walts and Walsh, 1988; Walsh et al., 1988), with the hydrogen donated by residue "Z" driving off the "R" group (as benzene from phenylmercury or methane from methylmercury), as the mercury associates with a second conserved cysteine residue (Fig. 9c --> d). Then the excess thiol compound required for enzyme activity (usually cysteine in vitro, but either glutathione or the enzyme mercuric reductase itself in vivo) removes the product Hg^{2+} from the lyase (Fig. 9d --> a). The turnover number for the organomercurial lyase is low, compared with the more rapid functioning of mercuric reductase (Begley et al., 1986b).

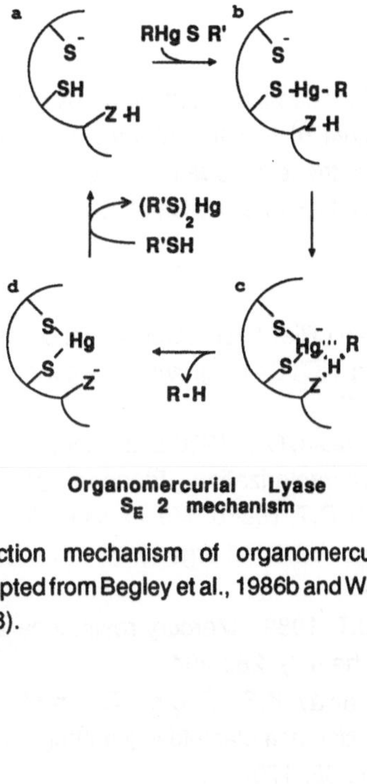

Organomercurial   Lyase
S_E 2 mechanism

FIG. 9 Reaction mechanism of organomercurial lyase (adapted from Begley et al., 1986b and Walsh et al., 1988).

SUMMARY

This relaxed "walk through" part of the Periodic Table of bacterial resistance mechanisms to toxic heavy metals (and their unfulfilled potential--sometimes-- for bioremediation) leads to several overall conclusions. Firstly, for essentially all toxic heavy metals that occur in the environment as cations or oxyanions, bacterial cells have evolved highly specific resistance mechanisms in order to survive in a frequently polluted world. The genes governing these resistance mechanisms frequently occur on plasmids and transposons, facilitating both molecular genetic studies in the laboratory and movement from cell to cell in the environment. Secondly, there are only a few overall mechanisms of resistance : (a) efflux pumping of the toxic material out from the cell; (b) bioaccumulation in a physiologically inaccessible form (either intracellular or extracellular), and (c) redox chemistry converting a more toxic ion species into one less toxic. Redox chemistry, although rare in bacterial cells, perhaps affords the most promising systems for bioremediation of toxic wastes. Reduction and volatilization of mercury from inorganic and organomercurials is the best understood of these resistance mechanisms. This process is ready for scale up and practical applications for aqueous mercury-polluted sites. Chromate reduction and arsenite oxidation are other such potential bioremediation systems, since these convert more toxic to less toxic forms. The biochemistry and molecular biology behind these two redox processes are not understood. Bioaccumulation also may be put to practical use. Although there is a sizable literature on initial efforts in preparing cellular bio-absorbers, commercial applications have not been accomplished beyond initial trials. Efflux pumping appears the least promising resistance mechanism for practical use, but even here there is the potential for building membranes with highly specific permeabilities to toxic salts. Until we understand the fundamental biochemical and genetic processes that bacterial cells use to achieve resistances to toxic metal ions, it is premature to rule out the possibility of these processes being useful in cleaning up environmentally polluted materials.

ACKNOWLEDGEMENTS

The work in our laboratory has been supported by grants from the National Science Foundation and National Institute of Health. Many people have contributed to 20 years progress. In more recent years, these include N.L. Brown, C. Hansen, D. and A. Nies, T.K. Misra, L. Chu, C. Cervantes, G. Nucifora, H. Ohtake, B.P. Rosen, and R.P. Novick.

REFERENCES

Aiking, H., H. Govers and J. van't Riet 1985. Detoxification of cadmium, mercury and lead in Klebsiella aerogenes NCTC418 growing in continuous culture. Appl. Environ. Microbiol. 50, 1262-1267.

Begley, T.P., Walts, A.E., and Walsh, C.T. 1986a. Bacterial organomercurial lyase: overproduction, isolation and characterization. Biochem. 25: 7186-7192.

Begley, T.P., Walts, A.E., Walsh, C.T. 1986b. Mechanistic studies of a protonolytic organomercurial cleaving enzyme: bacterial organomercurial lyase. Biochemistry 25: 7192-7200.

Belliveau, B.H., and Trevors, J.T. 1989. Mercury resistance and detoxification in bacteria. Appl. Organometallic Chem. 3: 283-294.

Bender, C.L., Malvick, D.K., Conway, K.E., George, S., and Pratt, P. 1990. Characterization of pXV10A, a copper - resistance plasmid in Xanthomonas campestris pv. vesicatoria. Appl. Environ. Microbiol. 56, 170-175.

Bopp, L.H. and Ehrlich, H.L. 1988. Chromate resistance and reduction in Pseudomonas fluorescens strain LB300. Arch. Microbiol. 150: 426-431.

Bopp, L.H. 1984. Microbial removal of chromate from contaminated waste water. U.S. patent # 4,468,461, issued August 28, 1984.

Brierley, C.L., Brierley, J.A., and Davidson, M.S. 1989. Applied microbial processes for metals recovery and removal from waste water. pp. 359-382. In T.J. Beveridge and R.J. Doyle (eds.) "Metal Ions and Bacteria", John Wiley & Sons, N.Y.

Brierley, J.A., Brierley, C.L., and Goyak, G.M. 1986. AMT-BIOCLAIM: a new waste water treatment and metal recovery technology. pp. 291-304. In "Fundamental and Applied Biohydrometallurgy" (R.W. Lawrence, R.M.R. Branion and H.G. Ebner, eds.) Elsevier, Amsterdam.

Brown, N.L. 1985. Bacterial resistance to mercury: reductio ad absurdum. Trends Biochem. Sci. 10: 400-403.

Cervantes, C., Ohtake, H., Chu, L., Misra, T.K., and Silver., S. 1990. Cloning, nucleotide sequence, and expression of the chromate resistance determinant of Pseudomonas aeruginosa plasmid pUM505. J. Bacteriol. 172: 287-291.

Cervantes, C. and Silver, S. 1990. Inorganic cation and anion transport systems of Pseudomonas. In "Pseudomonas: Biotransformations, Pathogenesis and Evolving Biotechnology" (eds. S. Silver, A.M. Chakrabarty, B. Iglewski, and S. Kaplan) American Society for Microbiology, Washington, D.C., pp. 359-372.

Chen, C.M., Misra, T.K., Silver, S. and Rosen, B.P. 1986. Nucleotide sequence of the structural genes for an anion pump. The plasmid-encoded arsenical resistance operon. J. Biol. Chem. 261: 15030-15038.

Cooksey, D.A. 1987. Characterization of a copper resistance plasmid conserved in copper-resistant strains of Pseudomonas syringae pv. tomato. Appl. Environ. Microbiol.53, 454-456.

Darnall, D.W. 1989. Removal and recovery of heavy metal ions from waste waters using a new bioabsorbant; AlgaSORB. In "Innovative Hazardous Waste Treatment Technology" (H. Freeman, ed.), Technomic Publishing company, Lancaster, PA, in press.

Darnall, D.W., Gabel, A.M., and Gardea-Torresday, J. !989. AlgaSORB: a new biotechnology for removing and recovering heavy metal ions from ground water and industrial waste water. pp. 113-124. In Hazardous Waste Treatment: Biosystems for Pollution Control, Proceedings of the 1989. A & WMA/EPA International Symposium. EPA, Cincinnati, Ohio.

Distefano, M.D., Au, K.G. and Walsh, C.T. 1989. Mutagenesis of the redox-active disulfide in mercuric ion reductase: catalysis by mutant enzymes restricted to flavin redox chemistry. Biochemistry 28: 1168-1183.

Distefano, M.D., Moore, M.J. and Walsh, C.T. 1990. Active site of mercuric reductase resides at the subunit interface and requires Cys135 and Cys140 from one subunit and Cys558 and Cys559 from the adjacent subunit: evidence from in vivo and in vitro heterodimer formation. Biochemistry 29: 2703-2713.

Dyke, K.G.H., Walters, J.A. and Curnock, S.P. 1991. Characterization of a staphylococcal plasmid that specifies resistance to cadmium ions., Manuscript in preparation.

Erardi, F.X., Failla, M.L. and Falkinham III, J.O. 1987. Plasmid-encoded copper resistance and precipitation by Mycobacterium scrofulaceum. Appl. Environ. Microbiol. 53: 1951-1954.

Gotz, F., Zebielski, J., Philipson, L. and Lindberg, M. 1983. DNA homology between the arsenate resistance plasmid pSX267 from Staphylococcus xylosus and the penicillinase plasmid pI258 from Staphylococcus aureus. Plasmid 9: 126-137.

Gvozdyak, P.I., Mogilevich, N.F., Ryl'skii, A.F. and Grishchenko, N.I. 1986. Reduction of hexavalent chromium by collection strains of bacteria. Mikrobiologiya 55: 962-965.

Hansen, C.L., Zwolinski, G., Martin, D. and Williams, J.W. (1984). Bacterial removal of mercury from sewage. Biotech. Bioengin. 26: 1330-1333.

Helmann, J.D., Ballard, B.T. and Walsh, C.T. 1990. The MerR metalloregulatory protein binds mercuric ion as a tricoordinate, metal-bridged dimer. Science 247: 946-948.

Helmann, J.D. and Walsh, C.T. 1990. Metal dependent transcriptional activation: binding of metal ions by the Bacillus species RC607 MerR protein. Unpublished Manuscript.

Horitsu, H., Futo, S., Miyazawa, Y., Ogai, S. and Kawai, K. 1987. Enzymatic reduction of hexavalent chromium by hexavalent chromium tolerant Pseudomonas ambigua G-1. Agric. Biol. Chem. 51: 2417-2420.

Hsu, C.M. and Rosen, B.P. 1989. Characterization of the catalytic subunit of an anion pump. J. Biol. Chem. 264: 17349-17354.

Hutchins, S.R., Davidson, M.S., Brierley, J.A. and Brierley, C.L. 1986. Microorganisms in reclamation of metals. Annu. Rev. Microbiol. 40: 311-336.

Ishibashi, Y., Cervantes, C. and Silver, S. 1990. Chromium reduction by Pseudomonas putida. Appl. Environ. Microbiol. 56: 2268-2270.

Karkaria, C.E. and Rosen, B.P. 1990. Mutagenesis of a nucleotide binding site of an anion-translocating ATPase. J. Biol. Chem. 265: 7832-7836.

Karplus, A. and Schulz, G.E. 1987. Refined structure of glutathione reductase at 1.54 Å resolution. J. Mol. Biol. 195: 701-729.

Khazaeli, M.B. and R.S. Mitra 1981. Cadmium-binding component in Escherichia coli during accomodation to low levels of this ion. Appl. Environ. Microbiol. 41: 46-50.

Komori, K., Wang, P.C., Toda, K. and Ohtake, H. 1989. Factors affecting chromate reduction in Enterobacter cloacae strain HO1. Appl. Microbiol. Biotechnol. 31: 567-570.

Komori, K., Rivas, A., Toda, K. and Ohtake, H. 1990a. Biological removal of toxic chromium using Enterobacter cloacae strain that reduces chromate under anaerobic conditions. Biotechnol. Bioengin. 35: 951-954.

Komori, K., Toda, K. and Ohtake, H. 1990b. Effects of oxygen stress on chromate reduction in Enterobacter cloacae. J. Ferment. Bioeng. 69: 67-69.

Komori, K., Rivas, A., Toda, K. and Ohtake, H. 1990c. A method for removal of toxic chromium using dialysis-sac cultures of a chromate-reducing strain of Enterobacter cloacae. Appl. Microbiol. Biotechnol. 33: 117-119.

Kusano, T., Ji, G., Inoue, C. and Silver, S. 1990. Constitutive synthesis of a transport function encoded by the Thiobacillus ferrooxidans merC gene cloned in Escherichia coli. J. Bacteriol. 172: 2688-2692.

Kvasnikov, E.I., Stepanyuk, V.V., Klyushnikova, T.M., Serpokrylov, N.S., Simonova, G.A., Kasatkina, T.P. and Pachenko, L.P. 1985. A new chromium-reducing, gram-variable bacterium with mixed type flagellation. Mikrobiologiya 54: 83-88.

Laddaga, R.A., Bessen, R. and Silver, S. 1985. Cadmium-resistant mutant of Bacillus subtilis 168 with reduced cadmium uptake. J. Bacteriol. 162, 1106-1110.

Lebedeva, E.V. and Lyalikova, N.N. 1979. Reduction of crocoite by Pseudomonas chromatophila sp. nov. Mikrobiologiya 48: 517-522.

Lee, B.T.O., Brown, N.L., Rogers, S., Bergemann, A., Camakaris, J. and Rouch, D.A., 1991. Bacterial response to copper in the environment: copper resistance in Escherichia coli as a model system. NATO ASI series vol. G23, pp. 625-632. In "Metal Specification in the Environment", J.A.C. Broekaert, S. Gucer, and F. Adams, eds. Springer Verleg, Berlin.

Lund, P.A. and Brown, N.L. 1989. Regulation of transcription from the mer and merR promoters of the transposon Tn501. J. Mol. Biol. 205: 343-353.

Meissner, P.S. and Falkinham III, J.O. 1984. Plasmid-encoded mercuric reductase in Mycobacterium scrofulaceum. App. Environ. Microbiol. 157: 669-672.

Mellano, M.A. and Cooksey, D.A. 1988. Nucleotide sequence and organization of copper resistance genes from Pseudomonas syringae pv. tomato. J. Bacteriol. 170: 2879-2883.

Miller, S.M., Moore, M.J., Massey, V., Williams, C.H. Jr., Distefano, M.D. ,Ballou, D.P., and Walsh, C.T. 1989. Two-electron reduced mercuric reductase binds Hg(II) to the active site dithiol but does not catalyze Hg(II) reductase. Biochemistry 28: 1194-1205.

Mobley, H.L.T. and Rosen, B.P. 1982. Energetics of plasmid-mediated arsenate resistance in Escherichia coli. Proc. Natl. Acad. Sci. USA 79: 6119-6122.

Moore, M.J. and Walsh, C.T. 1989. Mutagenesis of the N- and C-terminal cysteine pairs of Tn501 mercuric ion reductase: consequences for bacterial detoxification of mercurials. Biochemistry 28: 1183-1194.

Moore, M.J., Distefano, M.D., Walsh, C.T., Schliering, N. and Pai, E.F. 1989. Purification, crystallization, and preliminary x-ray diffraction studies of the flavoprotein mercuric ion reductase from Bacillus sp. strain RC607. J. Biol. Chem. 264: 14386-14388.

Nies, A., Nies, D.H. and Silver, S. 1989. Cloning and expression of plasmid genes encoding resistances to chromate and cobalt in Alcaligenes eutrophus. J. Bacteriol. 171: 5065-5070.

Nies, A., Nies, D.H. and Silver, S. 1990. Nucleotide sequence and expression of a plasmid-encoded chromate resistance determinant from Alcaligenes eutrophus. J. Biol. Chem. 265: 5648-5653.

Nies, D., Mergeay, M., Friedrich, B. and Schlegel, H.G. 1987. Cloning of plasmid genes encoding resistance to cadmium, zinc and cobalt in Alcaligenes eutrophus CH34. J. Bacteriol. 169: 4865-4868.

Nies, D.H., Nies, A., Chu, L. and Silver, S. 1989. Expression and nucleotide sequence of a plasmid-determined divalent cation efflux system from Alcaligenes eutrophus. Proc. Natl. Acad. Sci. USA 86: 7351-7355.

Nies, D.H. and Silver, S. 1989. Plasmid-determined inducible efflux is responsible for resistance to cadmium, zinc and cobalt in Alcaligenes eutrophus. J. Bacteriol. 171: 896-900.

Novick, R.P., Murphy, E., Gryczan, T.J., Baron, E. and Edelman, I. 1979. Penicillinase plasmids of Staphylococcus aureus: restriction-deletion maps. Plasmid 2: 109-129.

Nucifora, G., Chu, L., Silver, S. and Misra, T.K. 1989a. Mercury operon regulation by the merR gene of the organomercurial resistance system of plasmid pDU1358. J. Bacteriol. 171: 4241-4247.

Nucifora, G., Chu, L., Misra, T.K. and Silver, S. 1989b. Cadmium resistance of Staphylococcus aureus plasmid pI258 results from a Cd^{2+} efflux ATPase determined by the cadA gene. Proc. Natl. Acad. Sci. USA 86: 3544-3548.

O'Halloran, T.V. 1989. Metalloregulatory proteins: metal responsive molecular switches governing gene expression. vol. 25, pp. 105-145 In "Metal Ions in Biological Systems" H. Sigel, ed. Marcel Dekker, New York.

O'Halloran, T.V., Frantz, B., Shin, M.K., Ralston, D.M. and Wright, J.G. 1989. The merR heavy metal receptor mediates positive activation in a topologically novel transcription complex. Cell 56: 119-129.

Ohtake, H., Fujii, E. and Toda, T. 1990. Reduction of toxic chromate in an industrial effluent by use of a chromate-reducing strain of Enterobacter cloacae. Environ. Technol. Lett., 11: 663-668.

Ohtake, H., Cervantes, C. and Silver, S. 1987. Decreased chromate uptake in Pseudomonas fluorescens carrying a chromate resistance plasmid. J. Bacteriol. 169: 3853-3856.

Osborne, F.H. and Ehrlich, H.L. 1976. Oxidation of arsenite by a soil isolate of <u>Alcaligene</u> <u>s</u>. J. Appl. Bacteriol. <u>41</u>: 295-305.

Owolabi, J.B. and Rosen, B.P. 1990. Differential mRNA stability controls relative gene expression within the plasmid-encoded arsenical resistance operon. J. Bacteriol. <u>172</u>: 2367-2371.

Perry, R.D. and Silver, S. 1982. Cadmium and manganese transport in <u>Staphylococcus</u> <u>aureus</u> membrane vesicles. J. Bacteriol. <u>150</u>: 973-976.

Phillips, S.E. and Taylor. M.L. 1976. Oxidation of arsenite to arsenate by <u>Alcaligenes</u> <u>faecalis</u>. Appl. Environ. Microbiol. <u>32</u>: 392-399.

Ralston, R.M. and O'Halloran, T.V. 1990. Ultrasensitivity and heavy-metal selectivity of the allosterically modulated MerR transcription complex. Proc. Natl. Acad. Sci. USA <u>87</u>: 3846-3850.

Romanenco, V.I. and Kkoren'kov, V.N. 1977. A pure culture of bacteria utilizing chromates and bichromates as hydrogen acceptors in growth under anaerobic conditions. Mikrobiologiya <u>46</u>: 414-417.

Rosen, B.P., Weigel, U., Karkaria, C. and Gangola, P. 1988. Molecular characterization of an anion pump. The <u>arsA</u> gene product is an arsenite(antimonate)-stimulated ATPase. J. Biol. Chem. <u>263</u>: 3067-3070.

Rosenstein, R. and Götz, F. 1991. Nucleotide sequence and expression of arsenic resistance genes of <u>Staphylococcus</u> <u>xylosus</u>. Molec. Gen. Genet., Submitted.

Rouch, D., Camakaris, J., Lee, B.T.O. and Luke, R.K.J. 1985. Inducible plasmid-mediated copper resistance in <u>Escherichia</u> <u>coli</u>. J. Gen. Microbiol. <u>131</u>: 939-943.

Rouch, D., Lee, B.T.O. and Camakaris, J. 1989a. Genetic and molecular basis of copper resistance in <u>Escherichia</u> <u>coli</u>. pp. 439-446. In "Metal Ion Homeostasis: Molecular Biology and Chemistry" (eds. D.H. Hamer and D.R. Winge), Alan R. Liss, New York.

Rouch, D., Camakaris, J. and Lee, B.T.O. 1989b. Copper transport in <u>Escherichia</u> <u>coli</u>. pp. 469-477. In "Metal Ion Homeostasis: Molecular Biology and Chemistry" (eds. D. H. Hamer and D.R. Winge), Alan R. Liss, New York.

Sahlman, L., Lamier, A.-M., Lindskog, S. and Dunford, H.B. 1984. The reaction between NADPH and mercuric reductase from <u>Pseudomonas</u> <u>aeruginosa</u>. J. Biol. Chem. <u>259</u>: 12403-12408.

Sahlman, L., Lamier, A.-M. and Lindskog, S. 1986. Rapid-scan stopped-flow studies of the pH dependence of the reaction between mercuric reductase and NADPH. Eur. J. Biochem. <u>156</u>: 479-488.

Sandstrom, A. and Lindskog, S. 1987. Activation of mercuric reductase by the substrate NADPH. Eur. J. Biochem. <u>164</u>: 243-249.

San Francisco, M.J.D., Tisa, L.S. and Rosen, B.P. 1989. Identification of the membrane component of the anion pump encoded by the arsenical resistance operon of plasmid R773. Molec. Microbiol. <u>3</u>: 15-21.

San Francisco, M.J.D., Hope, C.L., Owolabi, J.B., Tisa, L.S. and Rosen, B.P. 1990. Identification of the metalloregulatory element of the plasmid-encoded arsenical resistance operon. Nucleic Acids Res. <u>18</u>: 619-624.

Shimada, K. and Matsushima, K. 1983. Isolation of potassium chromate-resistant bacterium and reduction of hexavalent chromium by the bacterium. Bull. Faculty Agriculture Mie Univ. <u>67</u>: 101-106.

Silver, S., Budd, K., Leahy, K.M., Shaw, W.V., Hammond, D., Novick, R.P., Willsky, G.R., Malamy, M.H. and Rosenberg, H 1981. Inducible plasmid-determined resistance to arsenate, arsenite and antimony(III) in <u>Escherichia</u> <u>coli</u> and <u>Staphylococcus</u> <u>aureus</u>. J. Bacteriol. <u>146</u>: 983-996.

Silver, S. and Keach, D. 1982. Energy-dependent arsenate efflux: the mechanism of plasmid-mediated resistance. Proc. Natl. Acad. Sci. USA 79: 6114-6118.

Silver, S. and Laddaga, R.A. 1990. Molecular genetics of heavy metal resistances in Staphylococcus plasmids. In "Molecular Biology of the Staphylococci" (R.P. Novick ed.), VCH Publishers, New York, pp. 531-549.

Silver, S. and Misra, T.K. 1988. Plasmid-mediated heavy metal resistances. Annu. Rev. Microbiol. 42: 717-743.

Silver, S., Nucifora, G., Chu, L. and Misra, T.K. 1989. Bacterial resistance ATPases: primary pumps for exporting toxic cations and anions. Trends Biochem. Sci. 14: 76-80.

Strandberg, G.W., Shumate II, S.E. and Parrott Jr., J.R. 1981. Microbial cells as biosorbents for heavy metals: accumulation of uranium by Saccharomyces cerevisiae and Pseudomonas aeruginosa. Appl. Environ. Microbiol. 41: 237-245.

Strandberg, G.W. and Arnold Jr., W.D. 1988. Microbial accumulation of neptunium. J. Indus. Microbiol. 3: 329-331.

Summers, A.O. and Silver, S. 1978. Microbial transformations of metals. Annu. Rev. Microbiol. 32: 637-672.

Tetaz, T.J. and Luke, R.K.J. 1983. Plasmid-controlled resistance to copper in Escherichia coli. J. Bacteriol. 154: 1263-1268.

Thieme, R., Pai, E.F., Schirmer, R.H. and Schulz, G.E. 1981. Three-dimensional structure of glutathione reductase at the 2 Å resolution. J. Mol. Biol. 152: 763-782.

Tisa, L.S. and Rosen, B.P. 1989. Molecular characterization of an anion pump: the ArsB protein is the membrane anchor for the ArsA protein. J. Biol. Chem. 265: 190-194.

Tisa, L.S. and Rosen, B.P. 1990. Transport systems encoded by bacterial plasmids. J. Bioenerg. Biomembr. 22: 493-507.

Trevor, J.T. 1987. Copper resistance in bacteria. Microbiol. Sci. 4: 29-31.

Tynecka, Z., Gos, Z., and Zajac, J. 1981. Energy-dependent efflux of cadmium coded by a plasmid resistance determinant in Staphylococcus aureus. J. Bacteriol. 147: 313-319.

Walsh, C.T., Distefano, M.D., Moore, M.J., Shewchuk, L.M. and Verdine, G.L. 1988. Molecular basis of bacterial resistance to organomercurial and inorganic mercuric salts. FASEB J. 2: 124-130.

Walts, A.E. and Walsh, C.T. 1988. Bacterial organomercurial lyase: novel enzymatic protonolysis of organostannanes. J. Amer. Chem. Soc. 110: 1950-1953.

Wang, P.C., Mori, T., Komori, K., Sasatsu, M., Toda, K. and Ohtake, H. 1989. Isolation and characterization of an Enterobacter cloacae strain that reduces hexavalent chromium under anaerobic conditions. Appl. Environ. Microbiol. 55: 1665-1669.

Wang, P.C., Mori, T., Toda, K. and Ohtake, H. 1989. Membrane-associated chromate reductase activity from Enterobacter cloacae. J. Bacteriol. 172: 1670-1672.

Weiss, A.A., Silver, S. and Kinscherf, T.G. 1978. Cation transport alteration associated with plasmid-determined resistance to cadmium in Staphylococcus aureus. Antimicrob. Agents Chemother. 14: 856-865.

Witte, W., Green, L., Misra, T.K. and Silver, S. 1986. Resistance to mercury and cadmium in chromosomally-resistant Staphylococcus aureus. Antimicrob. Agents Chemother. 29: 663-669.

Yoon, K.P. and Silver, S. 1991. A second gene in the cadA cadmium resistance determinant of Staphylococcus aureus plasmid pI258. J. Bacteriol., Submitted.

Yoon, K.P., Misra, T.K. and Silver, S. 1991. Regulation of the cadA cadmium resistance determinant of Staphylococcus aureus plasmid pI258. J. Bacteriol., Submitted.

THE POTENTIAL OF WHITE-ROT FUNGI IN BIOREMEDIATION

T. Kent Kirk[1] and Richard T. Lamar[1] and John A. Glaser[2]

[1]Forest Products Laboratory, U.S. Department of Agriculture
Forest Service, One Gifford Pinchot Drive, Madison, WI 53705, USA
[2]U.S. EPA Hazardous Waste and Engineering Research Laboratory
26 W. Martin Luther King Drive, Cincinnati, OH 45268, USA

ABSTRACT

The lignin-degrading enzyme system of white-rot fungi, which are mostly basidiomycetes, has been studied intensively in recent years. The extracellular component of the system is comprised of lignin peroxidase, manganese peroxidase, glyoxal oxidase and certain metabolites. Lignin is fragmented by this system, and the plethora of degradation products taken up by the hyphae and further metabolized by the intracellular system. The intracellular system has received very little research attention. The structural complexity and heterogeneity of lignin show in fact that this enzyme system is so nonspecific that it also degrades a variety of hazardous compounds, including polycyclic aromatics, some polychlorinated biphenyls and dioxins, DDT, and many chlorinated phenols. Using pentachlorophenol (PCP) as a model substrate, we have studied the possibility of using white-rot fungi to remediate soils contaminated with hazardous compounds. Successful laboratory results led to a field study in the summer of 1989. Results showed that the laboratory findings could be duplicated in the field.

INTRODUCTION

Until very recently, the literature concerning biodegradation and bioremediation of organic chemical wastes (xenobiotics) dealt almost exclusively with bacteria. It is now becoming apparent that fungi also play an important role in degrading organic materials in the ecosystem, and that they have potential for remediating contaminated soils and waters. The higher basidiomycetous fungi probably play the major role in recycling the carbon of lignocellulosics, which are the most abundant renewable organic materials on earth. Thousands of species of these fungi degrade the complex structural component lignin, thereby gaining access to the cellulose and hemicelluloses, which with lignin, make up lignocellulosics. During recent years, it has become clear that the lignin-degrading enzyme system of these fungi is quite nonspecific and catholic; many man-made as well as natural non-lignin compounds are degraded. This paper--and our research--concerns the deliberate harnessing of lignin-degrading fungi for remediating soils contaminated with hazardous organic chemi-

Biotechnology and Environmental Science: Molecular Approaches
Edited by S. Mongkolsuk et al., Plenum Press, New York, 1992

131

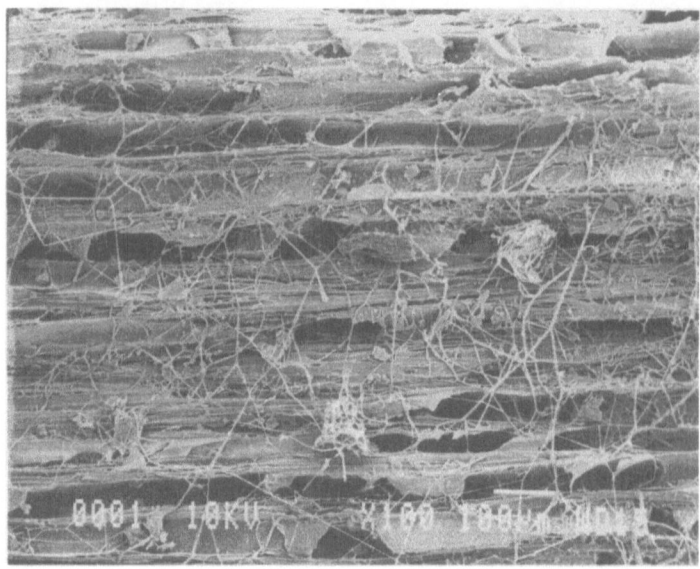

Figure 1. Scanning electron microscopic view of hyphae of *Phanerochaete chrysosporium* growing in cells of aspen (*Populus* spp.) wood; scale bar is 100 mm. (Courtesy of Dr. Irving Sachs)

cals. We describe the fungi, their ligninolytic system, and the compounds that they degrade. We then briefly summarize our laboratory and field studies with the wood preservative pentachlorophenol (PCP), which we have selected as a target chemical.

LIGNIN-DEGRADING FUNGI

Lignin-degrading fungi are ubiquitous. The most familiar are those that form mush-rooms, brackets (conks) and other sporophores on decaying trees, wood, forest litter, and other lignocellulosics. These are fungi that cause the white-rot type of wood decay, and the closely related litter-decomposing fungi. The most vigorous lignin-degraders are white-rot wood decay fungi, which are mainly basidiomycetes, and which in North America belong to the orders Agaricales, Aphyllophorales, and Tremellales. A few ascomycetes belonging to the order Speriales also cause white-rot wood decay. It is white-rot fungi that have been most intensively studied for bioremediation, and it is their lignin-degrading system that seems to be important in such applications. Figure 1 shows the hyphae of the most studied white-rot fungus, *Pheanerochaete chrysosporium*, growing in cells of aspen wood.

LIGNIN-DEGRADING SYSTEM OF WHITE-ROT FUNGI

The ligninolytic system of *P. chrysosporium* is illustrated schematically in Figure 2. The key extracellular enzymes are thought to be lignin peroxidase and glyoxal oxidase. The latter oxidizes the metabolites glyoxal and methyl glyoxal with reduction of O_2 to H_2O_2, which activates lignin peroxidase. Lignin peroxidase oxidizes nonphenolic aromatic nuclei in lignin by one electron to generate aryl cation radicals, which degrade nonenzymatically via many reactions. Most of those reactions result in polymer cleavages, generating both aromatic and aliphatic products which are taken up by the hyphae and mineralized. Little is known about the intracellular system. A second kind of peroxidase, manganese peroxidase, in the presence of H_2O_2, oxidizes Mn^{2+} to Mn^{3+}, which in turn can oxidize phenolic units in lignin. The role of such oxidation, if any, and that of manganese peroxidase, are not yet clear. The aromatic metabolite veratryl alcohol (Fig. 2) seems to play multiple roles, including stimulation of the production of the enzymes and electron transfer reactions during substrate degradation. For recent reviews of lignin degradation by *P. chrysosporium,* see Kirk (1988) and Eriksson et al. (1990). Both the extracellular and intracellular components of this system apparently plays roles in xenobiotic degradation.

XENOBIOTICS DEGRADED BY *P. chrysosporium*

Interest in using white-rot fungi for degrading hazardous chemicals originated from research on their use to decolorize kraft pulp bleach plant effluents. These effluents contain both polymeric, chlorinated, heavily oxidized fragments of lignin, which are responsible for the highly colored nature of the effluents, and a complex mixture of chlorinated phenols and other low molecular weight components. Several white-rot fungi, including *P. chrysosporium,* were found to decolorize the effluent (Eaton *et al.* 1982), and *P. chrysosporium* was shown to remove low molecular weight chloro-organics (Huynh *et al.* 1985). The results of these investigations led to further studies that demonstrated that *P. chrysosporium* and other fungi are able to degrade a broad range of structurally diverse xenobiotics (Table 1). The list of chemicals that are mineralized by white-rot fungi ranges from the insecticides DDT and Lindane, to wood-preserving chemicals, including PCP and the creosote components anthracene and phenanthrene, to polychlorinated biphenyls and dioxins.

Many of these compounds are substrates for lignin peroxidase, which presumably is involved in their initial degradation. For example, Hammel *et al.* (1986) reported that oxidation of anthracene by lignin peroxidase leads to anthraquinone, which is further metabolized by intact cultures. Other compounds mineralized by the cultures, including DDT and phenanthrene, are not substrates for lignin peroxidase. Pentachlorophenol is oxidized by lignin peroxidase with formation of tetrachloro-*p*-benzoquinone (Hammel and Tardone 1988), but whether this compound is an intermediate in its mineralization by *P. chrysosporium* is not yet known.

LABORATORY AND FIELD STUDIES OF PCP DEGRADATION IN SOIL BY *P. CHRYSO-SPORIUM*

Our studies of PCP degradation began with an investigation of the factors that influence growth of *P. chrysosporium* in three different soils (Lamar *et al.* 1987). Rank in terms of growth of *P. chrysosporium* in the three soils was Marshan > Zurich > Batavia. We found a positive correction between fungal growth and the soil nitrogen (N) and organic carbon (C) contents. Increasing the soil water potential from -0.03 MPa to -1.5 MPa resulted in greatly decreased growth of *P. chrysosporium,* indicating that the fungus prefers a fairly moist environment. Growth of the fungus was greater at 30°C and 39°C than at 25°C. Soil pH was without significant influence, presumably because the fungus controls the pH of its immediate environment. Under proper conditions, the fungus readily grows from wood chip inoculum into the surrounding soil (Fig. 3).

In further laboratory-scale studies we investigated the ability of white-rot fungi to

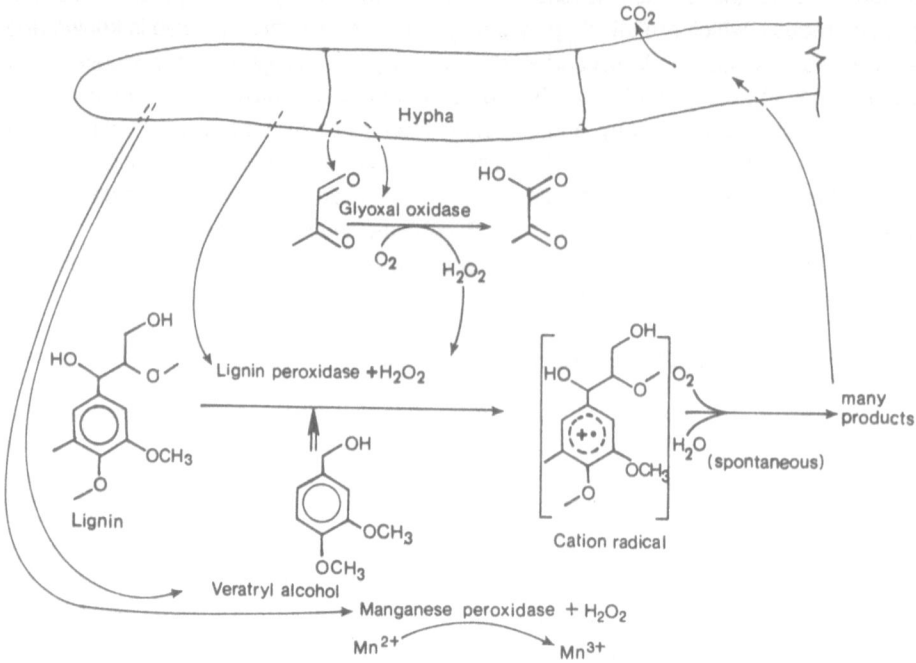

Figure 2. Schematic illustration of the ligninolytic system of *Phanerochaete chrysosporium*.

deplete PCP by following its fate in soils inoculated with various *Phanerochaete* spp. Inoculation of sterile Marshan, Zurich or Batavia soil with *P. chrysosporium*, which was one of the best degraders, resulted in a dramatic decrease (98%) in PCP concentration (Lamar *et al*. 1990). The rate of PCP depletion varied among the soils and appeared to relate to fungal growth and metabolic activity. Depletion of PCP by *P. chrysosporium* resulted mainly from its conversion to nonvolatile transformation products: loss of PCP via mineralization and volatilization was negligible. The nature of the transformation products--whether soil-bound or extractable--was greatly influenced by soil type. In the Marshan soil, *ca.* 60% of the PCP depletion was due to its conversion to extractable transformation products. Conversely, in the Batavia soil, 90% of the PCP depletion was due to its conversion to nonextractable products. Further study indicated that pentachloroanisole (PCA), is the primary extractable transformation product.

In the Fall of 1989, we conducted a field-scale study at a site contaminated by a commercial wood-preservative product that originally contained 84% mineral spirits, 1% paraffin wax, 10% alkyd varnish, and 5% technical grade PCP (4.3% PCP). Inoculation of the soil, which contained 250-400 mg g-1 PCP, with wood chips thoroughly colonized with either *P. chrysosporium* or *P. sordida* resulted in an overall decrease of 88%-91% of PCP in 6.5 weeks (Fig.4). This decrease was achieved under suboptimal temperatures for the growth and activity of the fungi. Comparison of the laboratory and field studies suggests that rates of depletion in field soils could be increased by controlling environmental conditions to favor fungal growth and activity.

In the field soil, 9% to 14% of the decrease in PCP was a result of methylation of PCP

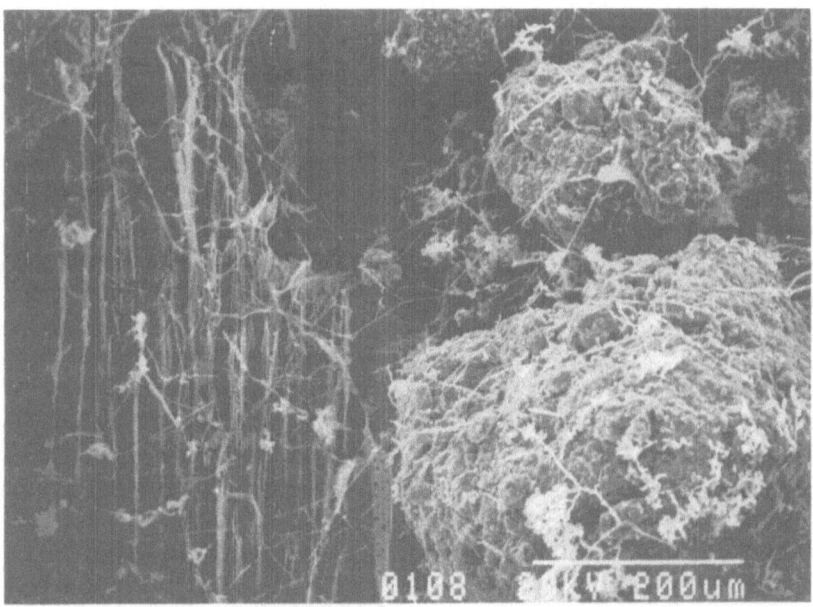

Figure 3. Hyphae of *Phanerochaete chrysosporium* growing from an inoculum wood chip into surrounding soil; scale bar is 200 mm. (Courtesy of Dr. M.J. Larsen.)

to PCA. Thus, methylation was not the major route of PCP depletion. Gas chromatographic analysis of sample extracts did not reveal the presence of extractable transformation products other than PCA. If loss of PCP via mineralization and volatilization was negligible, as in the laboratory studies, most of the PCP was converted to nonextractable soil-bound products. The nature of these products is not known. Bollag (1983) reported that chlorophenol-syringic acid hybrid polymers were produced when *Rhizoctonia praticola* laccase, a phenol-oxidizing enzyme, was exposed to syringic acid, a humus constituent, and chlorophenols. Similarly, oxidation of PCP in the field soil by ligninolytic enzymes of *P. chrysosporium* and *P. sordida* might have resulted in polymerization reactions, perhaps via quinonoid intermediates (Hammel and Tardone 1988), resulting in irreversible binding to organic matter.

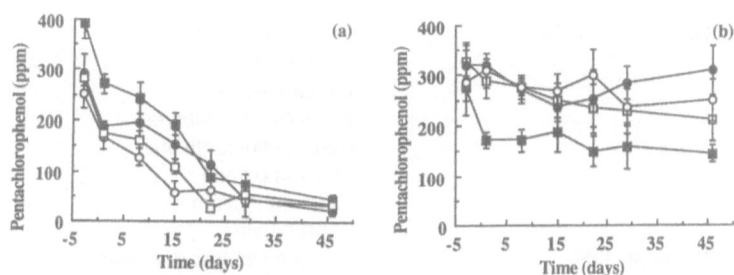

Figure 4. Concentration of PCP (mg g-1) in (a) soil inoculated with *Phanerochaete chrysosporium* (● ○) or *Phanerochaete sordida* (■□), or (b) soil receiving chips (■), chips and peat (□), peat (●) or no treatment (○).

The stability of xenobiotic-humic acid hybrid polymers under natural conditions is not known. However, work with artificially produced humic acid-xenobiotic hybrid polymers suggests that xenobiotics bound to humic materials through enzymatic polymerization reactions are relatively stable (Dec and Bollag 1985).

CONCLUSIONS

Based on the results of our investigations and those of others, we suggest that white-rot fungi have potential for use in the remediation of soils contaminated with hazardous compounds, including PCP. However, before use of these fungi can be considered a viable alternative, the nature, toxicity, and stability of the soil-bound products must be elucidated under a variety of conditions.

Table 1. Xenobiotics mineralized by white-rot fungi.

Reference	Xenobiotics Mineralized
Bumpus et al. 1985	1,1,1-trichloro-2,2-bis (4-chlorophenyl) ethane (DDT)
	Lindane
	2,3,7,8-TCDD
	3,4,3',4'-TCB
	Benzo(a)pyrene
Eaton 1985	Aroclor 1254
Arjmand and Sandermann 1986	4-Chloroaniline
Arjmand and Sandermann 1985	3,4-Dichloroaniline
	Chloroaniline-lignin conjugates
Haemmerli et al. 1986	Benzo(a)pyrene
Bumpus and Aust 1987	DDT
Kohler et al. 1988	
Mileski et al 1988	Pentachlorophenol
Lamar et al 1990	
Lin et al 1990	
Bumpus and Brock 1988	Triphenylmethane dyes
	Crystal violet
	Pararosaniline
	Cresol red
	Bromphenol blue
	Ethyl violet
	Malachite green
	Brilliant green
Ryan and Bumpus 1989	2,4,5-Trichlorophenoxyacetic acid
Bumpus 1989	Phenanthrene
Huttermann et al. 1989	Polycyclic aromatics
	Anthracene
	Fluorantherne
	Benzo(b)fluoranthene
	Benzo(k)fluoranthene
	Benzo(a)pyrene
	Indeno(ghi)pyrene
	Benzoperylen
Cripps et al. 1990	Azo and Heterocyclic dyes
	Orange II
	Tropaeolin O
	Congo Red
	Azure B
Fernando et al.1990	Trinitrotoluene

BIBLIOGRAPHY

Arjmand M. and H. Sandermann Jr.1985. Mineralization of chloroaniline/lignin conjugates and of free chloroanilines by the white rot fungus *Phanerochaete chrysosporium*. J. Ag. Food Chem. 33:1055-1060.

Arjmand M. and H.Sandermann Jr. 1986. Plant biochemistry of xenobiotics. Mineralization of cholroaniline/lignin metabolites from wheat by the white-rot fungus, *Phanerocha ete chrysosporium*. Z. Naturforschung 41c:206-214.

Bollag,J.M.1983. Cross-coupling of humus constituents and xenobiotic substances, pp. 127-141. In R.F. Christman and E.T. Gjessing (eds.), Aquatic and Terrestrial Humic Materials. Ann Arbor Science Publishers, Ann Arbor, Mich.

Bumpus J.A. 1989. Biodegradation of polycyclic aromatic hydrocarbons by *Phanerochaete chrysosporium*. Appl. Environ. Microbiol. 55: 154-158.

Bumpus, J.A. and B.J. Brock. 1988. Biodegradation of crystal violet by the white rot fungus *Phanerochaete chrysosporium*. Appl. Environ. Microbiol. 54: 1143-1150.

Bumpus J.A., M. Tien, D. Wright and S.D. Aust. 1985. Oxidation of persistent environmental pollutants by a white rot fungus. Science 228: 1434-1436.

Cripps, C., J.A. Bumpus and S.D. Aust. 1990. Biodegradation of azo and heterocyclic dyes by *Phanerochaete chrysosporium*. Appl. Environ. Microbiol. 56: 1114-1118.

Dec, J. and J.-M. Bollag. 1988. Microbial release and degradation of catechol and chloro-phenols bound to synthetic humic acid. Soil Sci. Soc. Am. J. 52: 1366-1371.

Eaton D.C. 1985. Mineralization of polychlorinated biphenyls by *Phanerochaete chrysosporium*, a ligninolytic fungus. Enz. Microbial Technol. 7: 194-196.

Eaton,D.C., H.-m. Chang, T.W. Joyce, T.W. Jeffries and T.K.Kirk. 1982. Method obtains fungal reduction of the color of extraction-stage kraft bleach effluents. Tappi J. 65: 89-92.

Eriksson, K.-E., R.A. Blanchette, and P.Ander. 1990. Microbial and Enzymatic Degradation of Wood and Wood Components. Springer Verlag, Berlin. 407 p.

Fernando, T., J.A. Bumpus and S.D. Aust. 1990. Biodegradation of TNT (2,4,6-trinitrotoluene) by *Phanerochaete chrysosporium*. Appl. Environ. Microbiol. 56: 1666-1671.

Hammel K.E. and P.J. Tardone. 1988. The oxidative 4-dechlorination of polychlorinated phenols is catalyzed by extracellular fungal lignin peroxidase. Biochemistry 27: 6563-6568.

Hammel K.E., B.Kalyanaraman and T.K.Kirk. 1986. Oxidation of polycyclic aromatic hydro-carbons and dibenzo[p]-dioxins by *Phanerochaete chrysosporium* ligninase J.Biol.Chem. 261: 16948-16952.

Haemmerli S.D.,M.S.A. Liesola, D.Sanglard and A.Feichter. 1986. Oxidation of benzo (a) pyrene by extracellular ligninases from *Phanerochaete chrysosporium*. J. Bio.Chem. 261: 6900-6903.

Huttermann, A.,J.Trojanowski and D. Loske. 1989. Process for the decomposition of complex aromatic substances in contaminated soils/refuse matter with micro-organisms. German Patent#DE3, 731, 816.

Huynh, V.-B.,H.-m. Chang, T.W.Joyce and T.K.Kirk. 1985. Dechlorination of chloro-organics by a white-rot fungus. Tappi J. 68: 98-102.

Kirk, T.K. 1988. Lignin degradation by *Phanerochaete chrysosporium*. ISI Atlas of Science: Biochemistry 1: 71-76.

Kohler A.,A.Jager, H. Willerhausen and H.Graf. 1988. Extracellular ligninase of *Phaneroch aete chrysosporium* Burdsall has no role in the degradation of DDT. Appl. Microbiol. Biotech. 29: 618-620.

Lamar, R.T., M.J.Larsen, T.K.Kirk, and J.A.Glaser. 1987. Growth of the white-rot fungus *Phanerochaete chrysosporium* in soil. pp. 419-424. In: Land Disposal, Remedial Action, Incineration and Treatment of Hazardous Waste. Proceedings of the Thirteenth Annual Research Symposium. May 6-8. U.S. Environmental Protection Agency, Cincinnati, Ohio.

Lamar, R.T., J.A. Glaser and T.K.Kirk. 1990. Fate of pentachlorophenol (PCP) in sterile soils inoculated with the white-rot basidiomycete *Phanerochaete chrysosporium*: mineralization, volatilization and depletion of PCP. Soil Biol. Biochem. 22: 433-440.

Lin, J.-E., H.Y. Wang and R.F.Hickey. 1990. Degradation Kinetics of pentachlorophenol by *Phanerochaete chrysosporium*. Biotech. Bioeng. 35: 1125-1134.

Mileski G.J.,J.A. Bumpus, M.A.Jurek and S.D.Aust. 1988. Biodegradation of pentachlorophenol by the white rot fungus *Phanerochaete chrysosporium*. Appl. Environ. Microbiol. 54: 2885-2889.

Ryan, T.P. and J.A. Bumpus. 1989. Biodegradation of 2,4,5-trichlorophenoxyacetic acid in liquid culture and in soil by the white rot fungus *Phanerochaete chrysosporium*. Appl. Microbiol. Biotech. 31: 302-307.

TOL PLASMID: ORGANIZATION AND REGULATION OF GENE EXPRESSION

Teruko Nakazawa, Masataka Tsuda, Sachiye Inouye* and Atsushi Nakazawa*

Department of Microbiology and *Department of Biochemistry, Yamaguchi University School of Medicine, Ube, Yamaguchi 755, Japan

The TOL plasmid of Pseudomonas putida mt-2 (ATCC 23973) specifies the complete degradation of toluene and xylenes. The aromatic hydrocarbons are oxidized through catechols and converted to central metabolites such as pyruvate and acetaldehyde (1).

CATABOLIC GENES ON TRANSPOSON

Analysis of the TOL plasmid was greatly facilitated by isolating RP4-TOL recombinants such as pTN2 (2). It was later shown that the TOL segment of pTN2 is a 56-kb transposon, Tn4651, which contains all the degrading genes and is able to transpose to various target replicons under the recA background (3). Tn4651 has Tn3-like properties with a transposase gene at one end and the resolution site as well as two trans-acting genes for cointegrate resolution at the other end.

It is now established that the degrading genes on the TOL plasmid are organized in two operons. The first, or the upper pathway operon OP1, consists of the promoter and structural genes for the enzymes involved in conversion of xylene to toluate. The second, or the lower operon OP2, consists of the promoter and 11 structural genes encoding enzymes for conversion of toluate to central metabolites (4). OP1 and OP2 are arranged in this order, and two regulatory genes, xylR and xylS, are located downstream from OP2. Transcription of xylR is in the same direction as OP2, whereas transcription of xylS is in the opposite direction.

CASCADE AMPLIFICATION IN GENE EXPRESSION

The current model for the regulation of gene expression of TOL plasmid includes cascade amplification (5). In the model, xylR plays a key role in the entire regulatory system whose expression is autogenously regulated. The xylR product, XylR, by combining with m-xylene, activates both the OP1 and the xylS promoters to initiate transcription. The over-produced xylS product, XylS, in turn activates OP2. Thus, the induction of the lower operon by the upper substrate involves a cascade response through amplification of the regulator. On the other hand, when m-toluate is present, XylS can activate OP2 even at a low basal level.

Biotechnology and Environmental Science: Molecular Approaches
Edited by S. Mongkolsuk et al., Plenum Press, New York, 1992

139

SIGMA 54-RNA POLYMERASE FOR TRANSCRIPTION OF xyl GENES

The xylR promoter has a similar structure to the E. coli canonical promoter of -35 and -10, which is recognized by the sigma 70-RNA polymerase (6), suggesting that P. putida also has a sigma 70-like RNA polymerase. The xy1R expression is autogenously regulated both in P. putida and E. coli.

On the other hand, the OP1 and the xy1S promoters are highly homologous with those of nitrogen-regulated or nitrogen fixation promoters at -24 and -12 which are dependent on RNA polymerase containing sigma 54 encoded by ntrA or rpoN.

The ntrA gene of P. putida was cloned and its nucleotide sequence was determined (7). The deduced 497-amino acid sequence is similar at some regions to those of sigma 54 of Azotobacter vinelandii, Klebsiella pneumoniae, and Rhizobium meliloti. The central conserved region is homologous to sigma 70 and RpoC of E. coli, which may play a role in interaction between sigma 54 and the core RNA polymerase. A helix-turn-helix motif for DNA binding is present near the C-terminal region. The N-terminal homologous region may be involved in interaction with XylR.

The ntrA mutants of P. putida were isolated which had pleiotropic defects (8, 9). In addition to xyl gene expression, utilization of nitrogen sources, amino acids, and C4-dicarboxylic acid as well as flagella formation are defective in the mutants.

xylR AND ENHANCER FOR TRANSCRIPTIONAL ACTIVATION

The xylR gene was cloned and its nucleotide sequence was determined (10). The deduced 566-amino acid sequence contains regions homologous to transcriptional activators of K. pneumoniae NtrC and NifA which are involved in expression of genes for nitrogen metabolism and nitrogen fixation genes in cooperation with sigma 54-RNA polymerase. The homologous C-terminal region has a helix-turn-helix motif which may interact with the upstream regulatory sequence of the target promoter. The central domain containing an ATP-binding motif which may be involved in the interaction with sigma 54-RNA polymerase. The N-terminal domain of XylR has no similarity to other transcriptional activators and may serve as an inducer-binding site.

The upstream regulatory sequence which is essential for XylR-dependent activation is found in the OP1 promoter ranging between -130 to -170 bp from the transcription start site (11). The sequence is active at longer distance (1.2 kb) and functions even after inversion. Thus the upstream regulatory sequence of OP1 has characteristic features of bacterial enhancers.

LOOPING MODEL FOR TRANSCRIPTIONAL ACTIVATION

As an enhancer-involving mechanism for transcriptional activation, a looping model rather than a sliding model is suggested based on the helical periodicity of the intervening DNA, i.e. insertion of full-turn length of DNA did not reduce efficiency of transcriptional activation, whereas half-turn insertion caused a drastic decrease (11).

In the model, the closed complex of sigma 54-RNA polymerase and the OP1 promoter is changed to form an open complex by XylR bound at the enhance sequence. For such interaction, sigma 54 and XylR should have a correct spacial relationship.

CONCLUSIONS

1. Catabolic genes are on a transposon. This may be the basis for evolution of the family of TOL plasmids by facilitating the gross transfer of substantial regions of genetic material.

2. Expression of the catabolic genes includes cascade amplification. The substrates of the upper catabolic pathway such as xylene and methylbenzyl alcohol are the initial signal. The signal is caught by a sensor XylR and transduced in two ways through transcriptional activation. One is to induce the catabolic enzymes to degrade substrates, and the other is to induce another regulatory protein, XylS, which in turn induces the lower pathway enzymes. By such amplification of the initial signal, the catabolic pathway functions efficiently to degrade the aromatic hydrocarbons otherwise toxic to the bacteria.

3. Both sigma 70 and sigma 54 are involved in the xyl gene expression. The xylR and OP2 promoters are transcribed by sigma 70-RNA polymerase, whereas the OP1 and the xylS promoters are transcribed by sigma 54-RNA polymerase. Sigma 54-RNA polymerase of P. putida may have evolved in bacteria for survival in the natural environments where preferable compounds are not available.

4. The upper operon and the xylS gene have upstream regulatory sequences located approximately 130 bp from the promoter which have characteristics of bacterial enhancers. A DNA looping model is suggested for transcriptional activation.

REFERENCES

1. Worsey, M.J., and P.A. Williams. 1975. Metabolism of toluene and xylenes by Pseudomonas putida (arvilla) mt-2: evidence for a new functions of the TOL plasmid. J. Bacteriol. 124: 7-13.
2. Nakazawa T., S. Inouye, and A. Nakazawa. 1980. Physical and functional mapping of RP4-TOL plasmid recombinations: analysis of insertion and deletion mutants. J. Bacteriol. 144: 222-231.
3. Tsuda, M., and T. Iino. 1987. Genetic analysis of a transposon carrying toluene degrading genes on a TOL plasmid pWWO. Mol. Gen. Genet. 210: 270-276.
4. Harayama, S., and M. Reikik. 1989. A simple procedure for transferring genes cloned in E. coli vectors into other Gram-negative bacteria: phenotypic analysis and mapping of TOL plasmid gene xylK. Gene 78: 19-27.
5. Inouye, S., A. Nakazawa, and T. Nakazawa. 1987. Expression of the regulatory gene xylS on the TOL plasmid is positively controlled by the xylR gene product. Proc. Natl. Acad. Sci. USA 84: 5182-5186.
6. Inouye, S., A. Nakazawa, and T. Nakazawa. 1985. Determination of the transcription initiation site and identification of the protein product of the regulatory gene xylR for xyl operons on the TOL plasmid. J. Bacteriol. 163: 863-869.
7. Inouye S., M. Yamada, A. Nakazawa, and T. Nakazawa. 1989. Cloning and sequence analysis of the ntrA (rpoN) gene of Pseudomonas putida. Gene 85: 145-152.
8. Inouye, S., A. Nakazawa, and T. Nakazawa. 1988. Nucleotide sequence of the regulatory gene xylR of the TOL plasmid from Pseudomonas putida. Gene 66: 301-306.

9. Köhler T., S. Harayama, J-L. Ramos, and K.N. Timmis. 1989. Involvement of *Pseudomonas putida* RpoN factor in the regulation of various metabolic functions. J. Bacteriol. 171: 4326-4333.

10. Inouye, S., M. Kimoto, A. Nakazawa. 1990. Presence of flagella in *Pseudomonas putida* is dependent on the *ntrA* (*rpoN*) gene. Mol. Gen. Genet. 221: 295-298.

11. Inouye, S., M. Gomada, U.M.X. Sangodkar, A. Nakazawa, and T. Nakazawa. 1990. Upstream regulatory sequence for transcriptional activator XylR in the first operon of xylene metabolism on the TOL plasmid. J. Mol. Biol. 216: 251-260.

MICROORGANISMS FOR THE INTRODUCTION OF CHLORINATED ALIPHATIC HYDROCARBONS INTO THE CARBON CYCLE

Thomas Leisinger

Mikrobiologisches Institut ETH, ETH-Zentrum
CH-8092 Zürich, Switzerland

Interest in microbial interactions with halogenated aliphatic hydrocarbons is due to the fact that these compounds have proven to be toxic environmental pollutants that are resistant to microbial degradation in many ecosystems. Their significance as pollutants has to do with the large quantities of these compounds produced worldwide (approx. 30 million tons per year) and with their widespread use as solvents or intermediates in chemical syntheses. The most important representatives, which are also the most significant pollutants, are trichloromethane, dichloromethane, tetrachloroethene, trichloroethene, chloroethene, trichloroethane and 1,2-dichloroethane. Halogenated aliphatic hydrocarbons enter the environment not exclusively as a result of industrial activities. Many of these compounds have been shown to occur naturally. They are synthesized by fungi, formed in forest fires and, most importantly, continuously released by marine macroalgae (Gschwend et al., 1985).

The three basic types of interactions of microorganisms with halogenated aliphatic hydrocarbons are listed in Table 1. These compounds may serve as the only carbon and energy sources for aerobic chemoheterotrophic bacteria, and the dehalogenation mechanisms involved are either substitutive or oxidative. Alternatively halogenated aliphatic hydrocarbons serve as cometabolic substrates for aerobic chemotrophs. In this case the organisms grow on another compound and fortuitously transform the halogenated compound to less halogenated or non-halogenated products by enzymes induced under the particular growth conditions. The dehalogenation mechanisms that have been observed under these conditions again are either substitutive or oxidative. As a third possibility, halogenated aliphatic hydrocarbons may function as electron acceptors in anaerobic systems thereby being reductively dehalogenated. This type of reaction has been observed in mixed (Vogel et al., 1987) and pure cultures (Egli et al., 1988) of strictly anaerobic bacteria. It is not known whether haloalkanes and haloalkenes are able to serve as physiological electron acceptors or whether electron transfer to these compounds is an accidental reaction without benefit to bacteria.

Biotechnology and Environmental Science: Molecular Approaches
Edited by S. Mongkolsuk et al., Plenum Press, New York, 1992

Table 1. Bacterial Systems for the Degradation of Aliphatic Halohydrocarbons

Function of Halohydrocarbon	Dehalogenation Mechanism
Carbon/energy source	substitutive oxidative
Cometabolic substrate	substitutive oxidative
Electron acceptor	reductive

SUBSTITUTIVE DEHALOGENATION

Several bacterial enzymes catalyzing the replacement of a halogen in haloalkanes by a water-derived hydroxyl group have been described. These haloalkane dehalogenases are specifically and exclusively involved in the utilization of halogenated aliphatic hydrocarbons as carbon sources for bacteria. As shown in Table 2 five enzymes of this type have been purified and characterized from one Gram-negative as well as from Gram-positive bacteria. All five enzymes are composed of a single subunit of roughly similar size. The enzymes differ with respect to inducibility, the *Xanthobacter* enzyme being formed constitutively whereas the other four enzymes are induced by haloalkanes. They also differ with respect to their substrate range. The preferred substrates are mono- or biterminally chlorinated, brominated or iodinated alkanes of varying chain length. The enzymes dehalogenate also the haloalcohols but very weakly or not at all the corresponding haloacids. The *Xanthobacter autotrophicus* haloalkane dehalogenase has been crystallized (Rozeboom et al., 1988). Its structural gene has been sequenced, and there was no indication of overall sequence similarity to other proteins (Janssen et al., 1989).

Table 2. Bacterial Haloalkane Dehalogenases

Organism	Subunits	MW (kDa)	Induction	Substrate range	Reference
Xanthobacter autotrophicus	1	36	-	C1-C4	(1)
Actinomycete-like	1	28	+	C2-C9	(2)
Corynebacterium sp.	1	36	+	C2-C12	(3)
Arthrobacter sp.	1	37	+	C1-C10	(4)
Rhodococcus sp.	1	34	+	C2-C16	(5)

(1) Keuning et al. (1985) (2) Janssen et al. (1988) (3) Yokota et al. (1987)
(4) Scholtz et al. (1987) (5) Sallis et al. (1990)

The carbon-halogen bond of dichloromethane is cleaved by a thiolytic reaction involving the sulfhydryl-group of glutathione as a nucleophile in substitutive dehalogenation. Facultatively methylotrophic bacteria of the genera *Hyphomicrobium*, *Methylobacterium* and *Pseudomonas* contain the enzyme dichloromethane dehalogenase which enables them to utilize CH_2Cl_2 as a carbon and energy source. This inducible enzyme, a hexamer with a subunit molecular mass of 37.5 kD, requires reduced glutathione as a cofactor. it catalyzes the thiolytic dehalogenation of CH_2Cl_2 by forming an S-chloromethyl-glutathione intermediate which undergoes hydrolysis to formaldehyde, reduced glutathione and hydrochloric acid (Kohler-Staub and Leisinger, 1985; Kohler-Staub et al., 1986):

$$CH_2Cl_2 + GSH \quad\quad ---> \quad\quad GS\text{-}CH_2Cl + HCl$$
$$GS\text{-}CH_2Cl + H_2O \quad ---> \quad\quad GS\text{-}CH_2OH + HCl$$
$$GS\text{-}CH_2OH \quad\quad\quad <---> \quad\quad GSH + CH_2O$$

This reaction is functionally similar to reactions catalyzed by eukaryotic glutathione S-transferases. Sequence analysis of the dichloromethane dehalogenase structural gene from *Methylobacterium* sp. DM4 has shown that this functional relationship is reflected at the structural level. Alignment of the deduced dichloromethane dehalogenase amino acid sequence with amino acid sequences of eukaryotic glutathione S-transferases revealed three regions containing highly conserved amino acid residues and indicated that the bacterial dichloromethane dehalogenase structural gene is a member of the glutathione S-transferase gene family (La Roche and Leisinger, 1990). Dichloromethane dehalogenases are strongly inducible by dichloromethane. In *Methylobacterium* sp. DM4 dehalogenase expression is subject to negative control by a trans-acting repressor protein encoded upstream of the dehalogenase structural gene.

OXIDATIVE DEHALOGENATION

Hydrolytic enzymes attacking chlorinated ethenes like chloroethene, trichloroethene and tetrachloroethene or polyhalogenated methanes like trichloromethane or tetrachloromethane have not been observed. To deal with these important environmental pollutants one has to rely on oxidative or reductive dehalogenation mechanisms. At present the biochemistry of these reactions is not understood in detail. Industrially important chlorinated aliphatic hydrocarbons which are utilized via oxidative dehalogenation reactions as growth substrates for bacterial pure cultures include chloromethane (Hartmans et al., 1986), 1,2-dichloroethane (Stucki et al., 1983) and chloroethene (Hartmans et al., 1985). However, many oxidative and all reductive dehalogenations of C1 and C2 halohydrocarbons are transformation reactions which do not provide the dehalogenating organisms with utilizable carbon sources.

Several important halogenated aliphatic hydrocarbons are co-oxidized by methane mono-oxygenase, an enzyme which is notorious for its wide substrate specificity (Oldenhuis et al., 1989; Tsien et al., 1989). In addition to methane monooxygenase, other oxygenases can initiate the degradation of halogenated aliphatic hydrocarbons. A well studied example is the dehalogenation of the important pollutant trichloroethene by the toluene dioxygenase of *Pseudomonas putida* strain F1 (Zylstra et al., 1989). Trichloroethene has also been reported to be decomposed by toluene monooxygenase (Winter et al., 1989), a propane monooxygenase (Wackett et al., 1989) and by the ammonia oxidizing bacterium *Nitrosomonas europaea* (Arciero et al., 1989). Trichloroethene decomposition catalyzed by toluene dioxygenase of *Pseudomonas putida* strain F1 has been examined at the genetic and enzymatic levels (Zylstra et al., 1989). Several transformation products of trichloroethene (e.g. protein, RNA, DNA) have been identified (Wackett and Householder, 1989). Since the toluene dioxygenase of *Pseudomonas putida* F1 is known to catalyze also monooxygenation reactions (Wackett et al., 1988), several degradation pathways are possible. However, from a chemical view, the formation of an epoxide-intermediate during degradation of trichloroethene seems to be most likely.

REDUCTIVE DEHALOGENATION

Xenobiotics with chain lengths of one or two carbon atoms and more than two halogens per molecule are susceptible to reductive dehalogenation. Due to the electronegative character of the halogen substituents, such compounds behave as electron acceptors. The halogenated solvents 1,1,1-trichloroethane, tetrachloromethane, trichloromethane and tetrachloroethene, are reductively dehalogenated in anaerobic sludge to less highly chlorinated compounds by different, largely unknown reactions (Vogel et al., 1987). Reductive dehalogenation of halogenated aliphatic hydrocarbons by pure cultures of anaerobic bacteria has been studied in several laboratories. The distribution of reductive dechlorination in the different organisms studied (Egli et al., 1990) leads us to correlate the acetyl-CoA pathway (Fuchs 1986) with dechlorination. This idea is supported by the data published on anaerobically dehalogenating pure cultures, all of which have this pathway (Belay and Daniels, 1987; Egli et al., 1987; Fathepure and Boyd, 1988). Bacteria with an operative acetyl-CoA pathway contain high levels of cobalt-corrinoids. Since these metalloporphyrins have been shown to mediate reductive dehalogenation of chlorinated C1-compounds *in vitro* (Krone et al., 1989), it appears likely that Co-corrinoids are the predominant catalysts for reductive dehalogenation in anaerobic bacteria. Metalloporphyrins are heat-stable, and we have observed that the rate of anaerobic degradation of tetrachloromethane was not affected by autoclaving cell suspensions of anaerobic bacteria (Egli et al., 1990). This supports the view that reductive dehalogenation is a non-enzymatic process, mediated by heat-stable cell components.

CONCLUSIONS

Bacteria for the partial and in many cases for the complete degradation of the important chlorinated aliphatic hydrocarbons are available. The most efficient systems are those, in which the chlorinated compounds are utilized aerobically as carbon sources via substitutive (hydrolytic, thiolytic) or oxidative dehalogenation mechanisms. Degradation rates in the range of 500 mg/g protein h^{-1} (e.g. dichloromethane, 1,2-dichloroethane) have been observed in these cases.

There is no single organism and no dehalogenation mechanism by which the whole range of chlorinated aliphatic hydrocarbons is dechlorinated. Major differences exist between the mechanisms occurring under aerobic conditions and those observed in anaerobic systems. Compounds that persist under aerobic conditions have been found to be degraded anaerobically. This is the case for several highly chlorinated solvents like tetrachloromethane, 1,1,1-trichloroethane and tetrachloroethene. These hazardous chlorinated hydrocarbons are reductively dehalogenated by anaerobic bacteria containing high levels of corrinoids. The most efficient anaerobic degradation (190 mg/g protein h^{-1}) was observed for tetrachloromethane. Pure cultures or syntrophic associations of anaerobic bacteria utilizing halohydrocarbons via initial hydrolytic dehalogenation as carbon and energy sources have not been described. They may exist for the utilization of some halohydrocarbons and would be of considerable interest in the treatment of contaminated groundwater.

With the exception of tetrachloromethane, the degradation rates for chlorinated aliphatic hydrocarbons are ten to hundred times lower than the degradation rates observed in aerobic systems. Reductive dehalogenation of highly chlorinated compounds occurs at progressively lower rates after each dehalogenation step. This leads to the accumulation of partially dehalogenated intermediates in anaerobic systems. Total degradation of highly chlorinated compounds like tetrachloroethene is feasible by combining anaerobic dehalogenation with subsequent treatment in an aerobic system.

REFERENCES

Arciero, D.,, Vannelli, T., Logan, M. Hooper, A.B. (1989) Biochem. Biophys. Res. Commun. 159, 640-643

Belay, N., Daniels, L. (1987) Appl. Environ. Microbiol. 53, 1604-1610

Egli, C., Scholtz, R., Cook, A.M., Leisinger, T. (1987) FEMS Microbiol. Lett. 43, 257-261

Egli, C., Tschan, T., Scholtz, R., Cook, A.M., Leisinger, T. (1988) Appl. Environ. Microbiol. 54, 2819-2824

Egli, C., Stromeyer, S., Cook, A.M., Leisinger, T. (1990) FEMS Microbiol. Lett. 68, 207-212

Fathepure, B.Z., Boyd, S.A. (1988) FEMS Microbiol. Lett. 49, 149-156

Fuchs, G. (1986) FEMS Microbiol. Rev. 39, 181-213

Gschwend, P.M., MacFarlane, J.K., Newman, K.A. (1985) Science 227, 1033-1035

Hartmans, S., de Bont, J.A.M., Tramper, J., Luyben, Luyben, K.C.A.M. (1985) Biotechnol. Lett. 7, 383-388

Hartmans, S., Schmuckle, A., Cook, A.M., Leisinger, T. (1986) J. Gen. Microbiol. 132, 1139-1142

Janssen, D.B., Gerritse, J., Brackman, J., Kalk, C., Jager, D., Witholt, B. (1988) Eur. J. Biochem. 171, 67-72

Janssen, D.B., Priens, F., van de Ploeg, J., Kazemier, B., Terpstta, P., Witholt, B. (1989) J. Bacteriol. 171, 6791-6799

Keuning, S., Janssen, D.B., Witholt, B. (1985) J. Bacteriol. 163, 635-639

Kohler-Staub, D., Leisinger, T. (1985) J. Bacteriol. 162, 676-681

Kohler-Staub, D., Hartmans, S., Gälli, R., Suter, F., Leisinger, T. (1986) J. Gen. Microbiol. 132, 2837-2843

Krone, U.E., Thauer, R.K., Hogenkamp, H.P.C. (1989) Biochemistry 28, 4908-4914

La Roche, S.D., Leisinger, T. (1990) J. Bacteriol. 172, 164-171

Oldenhuis, R., Vink, R.L.J.M., Janssen, D.B., Witholt, B. (1989) Appl. Environ. Microbiol. 55, 2819-2826

Rozeboom, H.J., Kingma, J., Janssen, D.B., Dijkstra, B. (1988) J. Mol. Biol. 200, 611-612

Sallis, P.J., Armfield, S.J., Bull, A.T., Hardman, D.J. (1990) J. Gen. Microbiol. 136, 115-120

Scholtz, R., Leisinger, T., Suter, F., Cook, A.M. (1987) J. Bacteriol. 169, 55016-5021

Stucki, G., Krebser, U., Leisinger, T. (1983) Experientia 39, 1271-1273

Tsien, H.C., Brusseau, G.A., Hanson, R.S., Wackett, L.P. (1989) Appl. Environ. Microbiol. 55, 3155-3161

Vogel, T.M., Criddle, C.S., McCarty, P.L. (1987) Environ. Sci. Technol. 21, 722-736

Wackett, L.P., Householder, S.R. (1989) Appl. Environ. Microbiol. 55, 2723-2725

Wackett, L.P., Kwart, L.D., Gibson, D.T. (1988) Biochemistry 27, 1360-1367

Wackett, L.P., Brusseau, G.A., Householder, S.R., Hanson, R.S. (1989) Appl. Environ. Microbiol. 55, 2960-2964

Winter, R.B., Kwang-Mu, Y., Ensley, B.D. (1989) Biotechnology 7, 282-285

Yokota, T., Omori, T., Kodama, T. (1987) J. Bacteriol. 169, 4049-40554

Zylstra, G.J., Wackett, L.P., Gibson, D.T. (1989) Appl. Environ. Microbiol. 5, 3162-3166

MOLECULAR EVOLUTION OF ENZYMES TOWARD

NYLON-OLIGOMERS, XENOBIOTIC COMPOUNDS

Hirosuke Okada, Seiji Negoro and Itaru Urabe

Department of Applied Microbial Technology,
Kumamoto Institute of Technology and
Department of Fermentation Technology, Osaka University, Japan

INTRODUCTION

The recent development of the chemical industry has brought about the distribution of a wide variety of synthetic compounds. Enzymes responsible for the degradation of synthetic compounds provide us with a suitable system for studying how microorganisms acquire such specific abilities.

Nylon-6 is produced by ring cleavage condensation of ε-caprolactam, and during the process, head-to-tail condensation of the intermediate mainly at the stages of dimer and tetramer results in the production of by-products of nylon factory, 6-aminohexanoate-cyclic-dimer and-tetramer, which amount about 5% of the raw material. We have isolated two bacterial strains, Flavobacterium sp. KI72 (KINOSHITA et al. 1975) and Pseudomonas sp. MK87 (KANAGAWA et al. 1989) as bacteria that can use 6-aminohexanoate-cyclic-dimer as the sole source of carbon and nitrogen.

NYLON-OLIGOMER DEGRADING ENZYMES

Three enzymes, 6-aminohexanoate-cyclic-dimer hydrolase [EC 3.5.2.12], (KINOSHITA et al. 1977; KANAGAWA et al. 1989) (F-EI for Flavobacterium KI72 and P-EI for Pseudomonas sp. NK87), and 6-aminohexanoate-dimer hydrolase [EC 3.5.1.46] (KINOSHITA et al. 1981) (F-EII for KI72 and P-EII for NK87) and 6-aminohexanoate-tetramer hydrolase [EC 3.5.2.-] (E-III) are responsible for the degradation of the cyclic oligomers. These enzymes were purified and characterized. Both F-EI and P-EI are homodimer enzymes with subunit of molecular weight of 51,000. They are active toward the cyclic dimer and not toward more than 100 kinds of natural amide bonds tested.

F-EII and P-EII are also dimer enzymes with two homologous subunits of molecular weight of 42,000. They are active on 6-aminohexanoate oligomers ranging from dimer to hexamer but not on icosamer and hectamer. We have tested more than 100 kinds of possible amide bonds of natural compounds but no activity was found.

Biotechnology and Environmental Science: Molecular Approaches
Edited by S. Mongkolsuk et al., Plenum Press, New York, 1992

149

F-EIII is either homodimer or trimer enzyme with subunit having molecular weight of 37,000. It is active on the cyclic-tetramer and pentamer and on the linear oligomers higher than trimer. EIII is not active on amide bonds of natural compounds tested.

From these results, we concluded the degradation sequence of the cyclic dimer is as follows.

$$NH-(CH_2)_5-CO \quad EI \quad H_2N-(CH_2)_5-CO \quad EII \quad H_2N-(CH_2)_5-COOH$$
$$CO-(CH_2)_5-NH \quad\quad\quad HOOC-(CH_2)_5-NH \quad\quad\quad \times 2$$

6-aminohexanoate-   6-aminohexanoate-   6-aminohexanoate
cyclic dimer   dimer

The unavailability of the natural amide bonds to be hydrolyzed by these enzymes might suggest that they created to adapt these xenobiotic compounds.

PLASMID DEPENDENCE OF EI, EII AND EIII

The wild strain of Flavobacterium sp. KI72 harbors three kinds of plasmid, pOAD1 (40kbp). pOAD2(46kbp) and pOAD3(56kbp). Curing and transformation experiments indicated that the genes coding EI, EII and EIII are located on pOAD2 (NEGORO et al. 1980). The restriction site (NEGORO and OKADA 1981) and the loci of nylA (coding EI), nylB (coding EII) and nylC (coding EIII) (NEGORO et al. 1983) are shown in Fig. 1. pOAD2 is hydrolyzed into 6 fragments by HindIII and are named from A to F fragments according to their size. nylA is located on the C fragment and nylB and nylC on the A fragment. These HindIII fragments were inserted into pBR322 and transformed to E. coli. The transformants harboring a hybrid plasmid containing the A fragment produced EII and EIII enzymes and those containing the C fragment produced EI enzyme.

Pseudomonas sp. NK87 harbors 6 kinds of plasmid; pNAD1 (20kbp), pNAD2 (23kbp), pNAD3 (51kbp), pNAD4 (57kbp), pNAD5 (76kbp) and pNAD6 (80kbp). Both the genes coding P-EI and P-EII were cloned into E. coli using pUC12 as the vector plasmid. These two enzymes were expressed in E. coli JM103. Using these cloned genes as the prove of Southern hybridization test, the genes for P-EI and P-EII were found to locate on pNAD2 and pNAD6 respectively. These results show the dependence of the genes on different plasmid in Pseudomonas sp. NK87, on the contrary to the case of Flavobacterium sp. KI72.

REPEATING SEQUENCES IN pOAD2

From the results of Southern DNA-DNA hybridization test, it was found that pOAD2 contained two kinds of repeating sequences (NEGORO et al. 1983). One is named RS1 which appears 5 times in this plasmid, might be responsible for the rearrangement of the plasmid in the same way that the insertion sequence and transposon elements are (CALOS and MILLER 1980). One piece of evidence for such activity of RS1 was obtained in the deletion plasmid pOAD21 from pOAD2, in which the deletion termini are located in or near the $RS1_D$ and $RS1_E$ region (Fig. 1). The other, RS2 appeared twice, one is in the A fragment and the other in the E fragment. RS2 in the A fragment is included in the nylB gene region suggesting a possibility that the RS2 region on the E fragment might be translocated to the RS2 on the A fragment to create a new enzyme (NEGORO et al 1984).

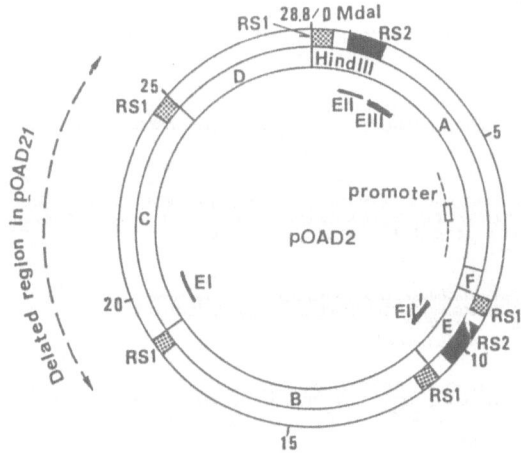

Fig. 1 Physical and functional map of pOAD2. Genes encoding EI, EII, EIII, and EII' are indicated. One type of repeating sequence(RS1) is shown as a shadow bar and the other repeated seuquence (RS2) as a solid bar.

EII AND ITS HOMOLOGOUS PROTEIN, EII'

The DNA sequence of two RS2 regions on the A and E fragments of pOAD2 are shown in Fig. 2 (OKADA et al. 1983). Fig 2. also shows the primary amino acid sequence deduced from the DNA sequence. The DNA base sequence is estimated to be the nylB gene from the amino acid sequence of N-terminal region determined with purified F-EII, the amino acid composition of the enzyme and the molecular weight. nylB gene has an open reading frame of 1176 bp which codes 392 amino acids. The DNA sequence of RS2 on the E fragment also contains an open reading frame able to encode a peptide of 392 amino acids (EII'). The gene coding EII' (nylB') shows 88% homology to nylB in the base sequence. Both nylB and nylB' have S.D. sequence, the initiation codon, and the termination codon at the same position. So, nylB' can code a protein of 392 amino acids which is also 88% homologous to EII, and 345 out of 392 amino acids are identical.

To produce EII' protein in E. coli, we inserted the 207 bp EcoRI fragment containing the lacUV5 promoter at the SmaI site located 218 bp upstream from the initiation codon of the nylB' gene (hybrid plasmid coding this gene was named pNDH10L1). In a double diffusion experiment, purified EII protein and the cell lysate of E. coli (pNDH10L1) produced precipitin lines to anti-EII serum, which fused at the corner with a spur. This result shows the immunological homology of the EII and EII' proteins. The EII' protein has 1/160 catalytic activity towards the linear dimer as EII at the basis of same antigenicity.

These results suggest that an ancestral nylB gene on pOAD2 was duplicated with the help of PS1 and one of them is mutated to increase the catalytic activity about 160 times.

151

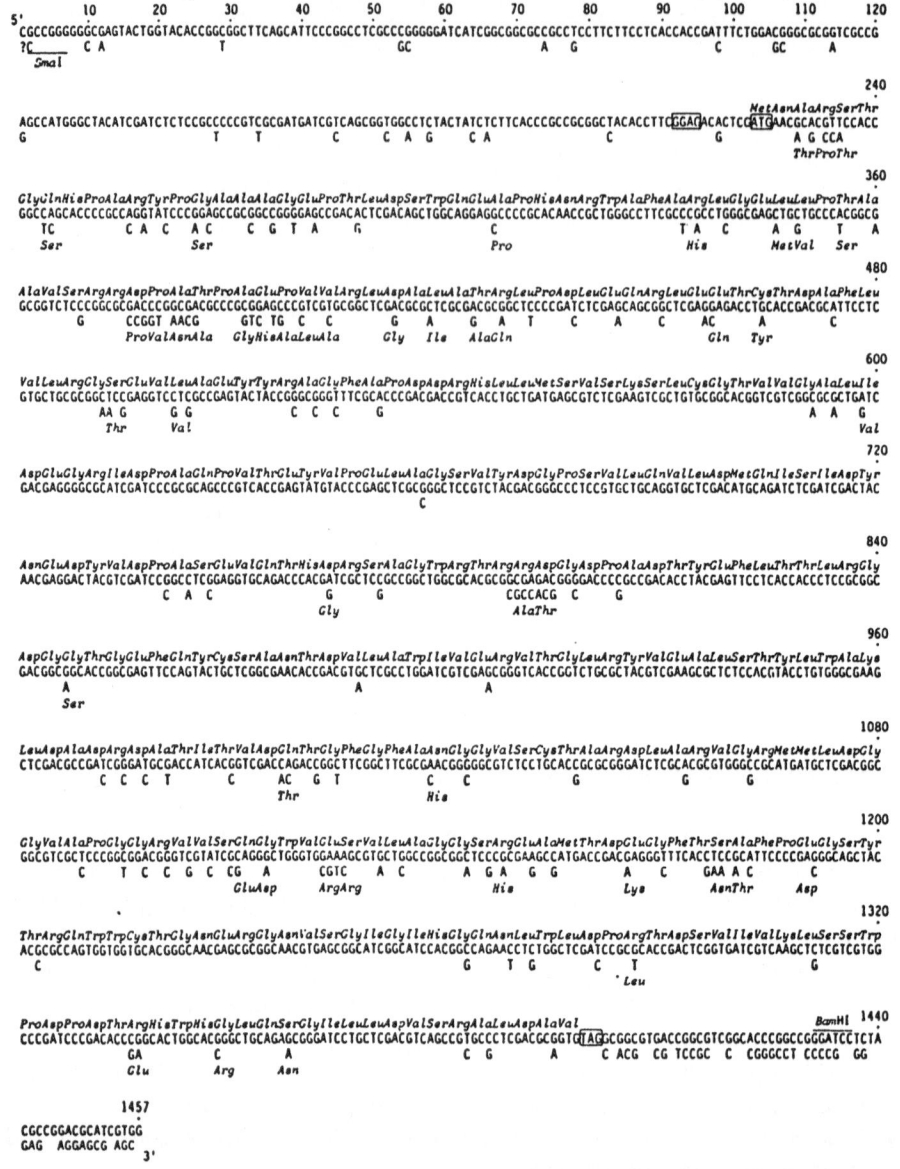

Fig. 2. Comparison of the sequence of nylB and nylB' and amino acid sequence deduced from them.

AMINO ACID ALTERATIONS CONTRIBUTED FOR THE ENZYME EVOLUTION (NEGORO et al. 1984)

Among 392 amino acids of EII and EII', only 47 are altered. To evaluate the contribution of the amino acid alterations to the increase of EII activity, we constructed chimera enzymes between EII and EII' (NEGORO et al. 1984). The restriction sites of PvuII, BgIII, SaII and BamHI (at respectively 74, 483, 771 and 1141 bp downstream from the initiation codon) are conserved in both nylB and nylB'. This allows us to exchange the DNA fragments at those restriction

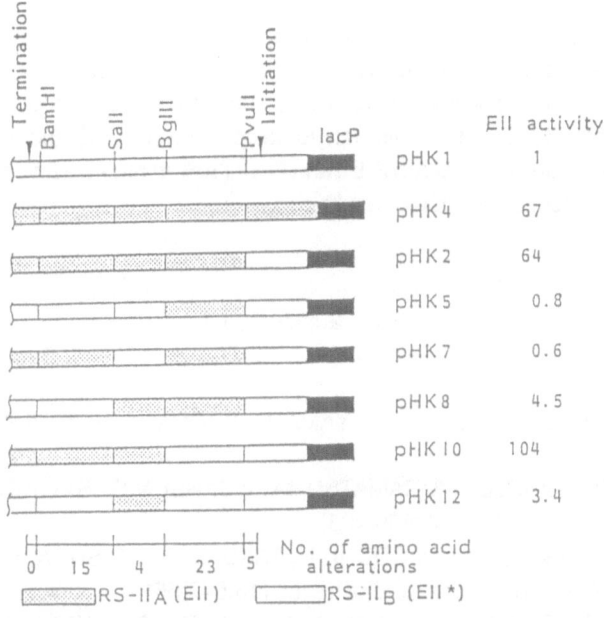

Fig. 3. Structure of chimera genes between nylB and nylB', and catalytic activity of the chimera enzymes.

sites between the two genes without changing the reading frame. The numbers of amino acids thus altered were 5 in the 74 bp fragment flanked by the initiation codon and the PvuII site (region I), 23 in the 409 bp fragment flanked by the PvuII and BglII sites (region II), 4 in the 288 bp fragment flanked by the BglII and SalI sites (region III) and 15 in the 370 bp fragment flanked by the SalI and BamHI sites (region IV). No amino acid alterations were encoded downstream from the BamHI site (region V). We constructed six intragenic recombinants coding chymera enzymes of EII and EII' as shown in Fig. 3. All chymera enzymes having region III originated from EII have more than 3 times activity as EII', and all chymera enzymes having region III and IV originated from EII have the activity of EII level, suggesting each one or more amino acid alterations encoded in the region III and IV stimulated the activity 160 times.

According to our unpublished results, the essential amino acid alteration for the evolution of EII which encoded in the region III is glycine[181] to aspartic acid, and the other amino acid alteration in region IV is histidine[266] to asparagine. This conclusion was confirmed by constructing a mutant EII' enzyme with mutation at Gly[181] to Asp and His[266] to Asn having similar catalytic activity as EII. These results indicate that very few amino acid alterations are required for an enzyme evolution such as evolution of EII from EII'.

HOMOLOGY BETWEEN F-EI AND P-EI (TSUCHIYA et al. 1989)

Antiserum against the purified F-EI enzyme made a precipitine line with cell extracts of NK87, which fused at the corner to the precipitine line formed by the F-EI enzyme, and the

^{32}P-labeled F-nylA probe hybridized to DNA fragments containing the -nylA gene. Both nylA genes were cloned and sequenced. Open reading frames of 1,479 bp starting at a GTG and terminating at TGA codon were found in both genes. Very high homology of more than 99% was observed. 493 amino acid sequences of F-EI and P-EI both deduced from the DNA base sequences differ only 7 out of 493. This high homology between P-EI and F-EI makes a good contrast to the low homology between P-EII and F-EII (35% in amino acid sequence and 53% in DNA base sequence; unpublished results).

LITERATURE

CALOS, M.P., MILLER, J.M. 1980 Cell 20:579-595.

KANAGAWA., K., NEGORO, S., TAKADA, H., OKADA, H. 1989 J. Bacteriol. 171:3181-3189.

KINOSHITA, S., KAGEYAMA, S., IBA, K., YAMADA, Y., OKADA, H. 1975 Agric. Biol. Chem. 39:1219-1223.

KINOSHITA, S., NEGORO, S., MURAMATSU, M., VISARIA, V.S., SAWADA,S., OKADA, H. 1977 Eur. J. Biochem. 80:489-495.

KINOSHITA, S., TERADA, T., TANIGUCHI, T., TAKENE, Y., MASUDA, S., MATSUNAGA, N., OKADA, H. 1980 Eur. J. Biochem. 116:547-551.

NEGORO, S., SHINAGAWA, H., NAKATA, A., KINOSHITA, S., HATOZAKI, T., OKADA, H. 1980 J. Bacteriol. 143:238-245.

NEGORO, S., TANIGUCHI, T., KANAOKA, M., KIMURA, H., OKADA, H. 1983 J. Bacteriol. 155:22-31.

NEGORO, S., NAKAMURA, S., OKADA, H. 1984 J. Bacteriol. 158:419-424.

NEGORO, S., NAKAMURA, KIMURA, H., FIJIYAMA, K., ZHANG, Y., KANZAKI, N., OKADA, H. 1984 J. Biol. Chem. 259:13648-13651.

NEGORO, S., MITAMURA, T., OKA, K., KANAGAWA, K., OKADA, H. 1989 Eur. J. Biochem. 185:521-524.

OKADA, H., NEGORO, S., KIMURA, H., NAKAMURA, S. 1983 Nature(London) 206:203-206.

TSUCHIYA, K., FUKUYAMA, S., KANZAKI, N., KANAGAWA, K., NEGORO, S., OKADA, H. 1989 J. Bacteriol. 171:3187-3191.

RESPONSES OF E. coli TO DNA DAMAGE AND STRESS

Toshihiro Ohta, John R. Battista, Caroline E. Donnelly, and
Graham C. Walker

Department of Biology, Massachusetts Institute of Technology
Cambridge MA 01239

ABSTRACT

Exposure of *Escherichia coli* to agents that damage DNA or interfere with DNA
replication results in the induction of the SOS response. A number of chromosomal genes
that are repressed by the LexA protein are transcribed at higher levels and various lysogenic
bacteriophage are induced. The RecA protein becomes activated by binding to some
intracellular inducing signal, probably single-stranded DNA and then mediates proteolytic
cleavage of LexA and bacteriophage repressors by facilitating an otherwise latent capacity
of these molecules to autodigest. The products of the SOS-regulated operon *umuDC* are
required for most UV and chemical mutagenesis. We have shown that the UmuD protein
shares homology with the carboxyl-terminal domains of LexA and several bacteriophage
repressors and is activated for its role in mutagenesis by a RecA-mediated proteolytic event.
Thus the regulation of *umuD* involves a transcriptional derepression and a posttranslational
activation that are mechanistically and evolutionarily related. A set of missense mutants of
umuD was isolated and shown to encode mutant UmuD proteins that are deficient in RecA-
mediated cleavage *in vivo* but which can be partially cleaved at a higher UV dose. Most of
these mutations are dominant to *umuD*+ with respect to UV mutagenesis yet do not interfere
with SOS induction. Although both UmuD and UmuD' form homodimers, we have found
evidence that they preferentially form heterodimers. These studies of *umuD* have suggested
a role for intact UmuD in the modulation of SOS mutagenesis. Other genetic studies have
indicated that the RecA protein plays a third role in mutagenesis besides mediating the
cleavage of LexA and UmuD. In addition, we have observed that efficiency of UV mutagenesis
is greatly reduced by mutations affecting the *groES* and *groEL* heat-shock genes. These genes
encode proteins that function as molecular chaperones which mediate protein folding and
protein-protein interaction. It seems possible that they may play a role in the proper assembly
of a protein complex required for SOS mutagenesis.

THE umuDC LOCUS OF E. coli AND ITS ANALOGS

Despite many years of study, the molecular mechanism of mutagenesis in *Escheric*

Biotechnology and Environmental Science: Molecular Approaches
Edited by S. Mongkolsuk. et al., Plenum Press, New York, 1992

155

hia coli has not yet been determined nor has it been determined in any other organism. Experiments of Weigle (53) provided the first clue that, in *E. coli* at least, the process of UV mutagenesis requires the participation of host functions and that the expression of one or more of these functions is inducible. By screening for *E. coli* mutants that were nonmutable by UV and various agents, Kato and Shinoura (27) and Steinborn (48) independently identified a locus, *umuC*, whose function was required for most mutagenesis by UV and many chemical mutagens. In earlier work, we showed that this locus consists of two genes, *umuD* and *umuC*, organized in an operon (16). We have also characterized an evolutionary-diverged but functionally-equivalent analog of this operon, *mucAB* (40, 41), that is present on the plasmid pKM101 (36). Sequencing of the *umuDC* operon (28, 40) and of the *mucAB* operon (40) revealed that the deduced amino acid sequences of the UmuD and MucA proteins are 41% identical and those of the UmuC and MucB proteins are 55% identical. Strike et al. (49, 50; personal communication) have since determined the sequence of the plasmid TP110-encoded *impAB* operon whose proteins are also functionally analogous to Umu proteins and approximately 50% related by sequence to both the Umu and Muc proteins. Interestingly, Sedgwick et al. (45) have obtained evidence that the *umuDC* genes of *E. coli* are, or were, located on a transposable genetic element. At this point it seems quite likely that those procaryotes that are mutable by UV (51) encode *umuDC*-like functions, that those that are nonmutable by UV (51) lack *umuDC*-like functions, and that there are *umuDC* analogs on a number of naturally occurring plasmids (38).

RELATIONSHIP OF THE SOS RESPONSE TO umuDC FUNCTION AND TO MUT-AGENESIS

The regulation of the *umuDC* operon and the *mucAB* operon is intimately intertwined with the regulation of the SOS regulon of *E. coli* both at the level of transcriptional control and at the level of posttranslational activation. We have shown that, like other SOS-regulated loci, the *umuDC* and *mucAB* operons are repressed by the LexA protein (1, 17). When RecA becomes activated after an SOS-inducing treatment (51,52), it mediates the proteolytic cleavage of the LexA molecule at its Ala[84]-Gly[85] bond (51, 52). Recent work of Sassanfar and Roberts (43) has indicated that the signal for RecA activation is the production of single-stranded DNA gaps when replication stops at a damage-induced lesion. Little has found that purified LexA cleaves at the same Ala-Gly bond when incubated at pH 9-10 thereby suggesting that activated RecA (RecA') acts by facilitating an otherwise latent capacity of LexA to autodigest (31). The cleavage of LexA leads to increased transcription of *umuDC* and other SOS-regulated loci (51, 52).

LexA shares homology with the repressors of bacteriophages, 1, 434, P22, and φ80 and cleavage of these proteins appears to occur by an analogous mechanism (15, 44). The cleavage site of all these proteins is an Ala-Gly bond except for φ80 repressor which has a Cys-Gly cleavage site (15). Slilaty and Little (47) have recently suggested that the hydrolysis of the LexA Ala[84]-Gly[85] bond proceeds by a mechanism similar to that of serine proteases, with Ser[119] acting as a nucleophile and Lys156 as an activator. In support of this model they reported that Ser[119]→Ala and Lys[156]→Ala changes yield functional LexA derivatives that do not cleave in the presence of activated RecA and do not undergo autodigestion at alkaline pH. In contrast, they found that changing Ser[119] to Cys, which is nucleophilic, yield a LexA derivative that undergoes substantial cleavage.

When we sequenced the *umuDC* and *mucAB* operons, we noted that UmuD and MucA shared amino acid similarly with the carboxyl-terminal domain of LexA and the family of phage repressors (40) (Fig. 1). On the basis of that observation, we proposed the hypothesis that UmuD is posttranslationally activated by cleavage at its Cys[24]-Gly[25] bond by activated RecA and that the amino terminal polypeptide of UmuD was expendable (40). This hypothesis has now been proven. Shinagawa et al. (46) used antibodies to UmuD to demonstrate RecA-mediated cleavage of UmuD in vivo, Burckhardt et al. (10) used purified UmuD to demonstrate Rec-A mediated cleavage and autodigestion at alkaline pH in vitro, and my lab (39) has carried out a series of genetic experiments showing that this cleavage activates UmuD for its role in mutagenesis.

RecA-MEDIATED CLEAVAGE ACTIVATES UmuD

By introducing overlapping termination (TGA) and initiation (ATG) codon at the site in the umuD sequence that corresponds to the cleavage site, we produced an engineered umuD gene that encodes two polypeptides that are virtually the same (39) as those produced normally only by RecA-mediated cleavage of the UmuD protein. This engineered umuD gene complemented the deficiencies of a umuD mutant in mutagenesis as efficiently as a umuD+ gene thereby ruling out the possibility that the cleavage inactivated UmuD. By subsequent site-directed mutagenesis experiments we were able to show that the 12 kDa carboxyl-terminal domain of UmuD (termed UmuD') was both necessary and sufficient for its role in UV mutagenesis. This conclusion was complemented by a set of site-directed mutations that altered the cleavage site and certain conserved residues of UmuD largely blocked RecA-mediated UmuD cleavage and reduced UV mutagenesis to approximately 3-15% of that observed with wild type UmuD. We demonstrated (39) that the purpose of RecA-mediated cleavage of UmuD is to activate UmuD for its role in mutagenesis by showing that plasmids encoding either the engineered form of UmuD encoding two polypeptides or the carboxyl-terminal polypeptide of UmuD (i.e. UmuD') could restore UV-mutability to a lexA(Def) recA430 strain while plasmids encoding umuD+ could not (39). (The RecA430 protein is able to participate in recombination but is partially defective in its ability to cleave LexA and completely defective in its ability to cleave UmuD; recA430 and lexA(Def) recA430 strains are UV nonmutable.) Thus RecA-mediated cleavage of UmuD represents a new dimension to the SOS response in which a posttranslation event mechanistically related to a transcription depression event.

We also used site-directed mutagenesis techniques (39) to test further the hypothesis of Slilaty and Little (47) that the mechanism of cleavage of LexA and phage repressors is related to that of serine proteases (25). Our observations that Ser[60]→Ala and Lys[97]→Ala blocked UmuD cleavage and greatly reduced UV mutagenesis whereas a Ser[60]→Cys changes only partially reduced mutagenesis were consistent with the effect of the corresponding mutations in LexA (47). By introducing the same mutations into an engineered umuD gene encoding UmuD' we were able to show that the primary role of Ser[60] is to mediate the cleavage of UmuD whereas Lys[97] is also necessary for the subsequent role of activated UmuD in mutagenesis (39).

A THIRD ROLE FOR RecA IN MUTAGENESIS

In the course of genetically analyzing the significance of the cleavage of UmuD, we uncovered evidence suggesting that RecA plays a third role in mutagenesis besides mediating the cleavage of LexA (which leads to increased transcription of umuDC) and mediating the cleavage of UmuD (which activates UmuD for it role in mutagenesis) (39). A collaborative study of other *recA* alleles provided further support for this hypothesis (14). Ennis *et al.* (18) also carried out a detailed analysis of the properties of a set of *recA* mutants and independently reached the same conclusion. This third role for RecA in mutagenesis is not yet known although *in vitro* studies have shown that RecA can inhibit the 3'-5' proofreading exonuclease of the epsilon subunit of DNA polymerase III (19, 33).

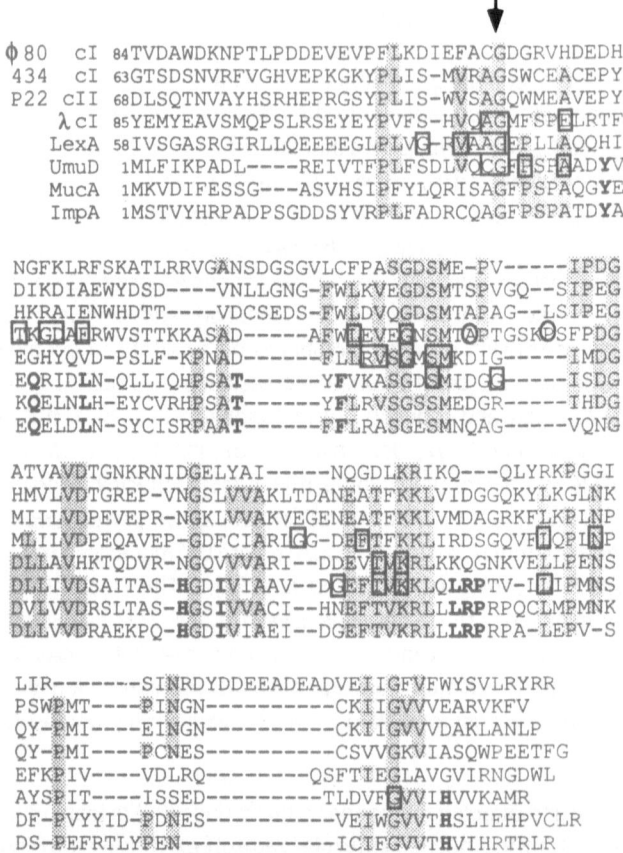

Fig. 1 Homology between the bacteriophage ø80, 434, P22, and l repressors, LexA, and the mutagenesis proteins UmuD, MucA, and ImpA (5). Amino acids that are identical between four or more members of the set are shaded. Positions of l (21), LexA (30), and UmuD (5, 39) where amino acid substitutions have been shown to yield stable proteins that are defective in RecA-mediated cleavage are indicated by squares. Positions of l repressor where an amino acid substitution has been shown to interfere with dimer formation are indicated by circles (22). Amino acids that are identical in the three mutagenesis proteins but are not shared with LexA or the bacteriophage repressors are indicated by bold lettering. The cleavage site is indicated by an arrow. The ImpA sequence is reproduced with the permission of P. Strike (University of Liverpool) (personal communication).

POSSIBLE ROLES FOR INTACT UmuD IN MODULATING SOS MUTAGENESIS.

In a recent study, we isolated a set of missense mutants of *umuD* and showed that these encode stable mutant UmuD proteins that are deficient in RecA-mediated cleavage *in vivo* (5). The amino acid changes caused by these mutations are: CY24, GS25, GD25, PS27, AT30, GR65, GD92, TM95, LM107, and GD129. Most of these mutations are dominant to *umuD⁺* with respect to UV mutagenesis yet do not interfere with SOS induction. In the course of considering possible mechanisms to account for the dominance of the *umuD* mutants, we made the discovery that, although both UmuD and UmuD' form homodimers, they preferentially form heterodimers. The results obtained in this study have led us to consider the hypothesis that intact UmuD is not simply an inactive form of UmuD' but is rather a dominant inhibitor of UmuD'-dependent mutagenesis. The dominance of the *umuD* alleles that reduce RecA-mediated cleavage and the preferential generation of UmuD-UmuD' heterodimers are observations consistent with this concept. UmuD is potentially well-suited to a negative regulatory role. As the SOS response begins to shut off, an accumulation of UmuD could lead to the formation of UmuD-UmuD- heterodimers and hence to an inhibition of UmuD' activity. UmuD is cleaved much less efficiently than LexA both *in vivo* (46) and *in vitro* (10). Thus one would expect some intact UmuD to accumulate before the increase in intact LexA would return expression of the *umuDC* operon to its basal level. Since the results of our *in vitro* experiments suggest that the formation of heterodimers is favored over homodimers, heterodimers should begin to form as soon as intact UmuD begins to accumulate.

The results of several previous studies, when considered collectively, suggest to us that there is indeed a mechanism for the deactivation of SOS-induced mutagenic capability that could be accounted for by the model we have proposed. Using two different approaches, Witkin (55) and Defais *et al.* (12) showed that SOS mutagenesis decays with a half-life of approximately 30 min in a Uvr background. However, Sassanfar and Roberts (43) have recently shown that, in Uvr⁻ strain irradiated with an even lower dose of UV than that used by Defais *et al.* (12), LexA continues to be cleaved at a maximal rate for at least 60 min after the exposure to UV. Thus the decay of SOS mutagenesis cannot be accounted for by the disappearance of RecA*, the accumulation of LexA, and the subsequent repression of the SOS regulon. Furthermore, pulse-chase studies (Battista and Walker, unpublished results), have indicated that the UmuD' and UmuC proteins are stable for at least two hours following translation, apparently ruling out proteolytic degradation of these proteins as a mechanism for inactivating SOS-induced mutagenic potential within the time frames discussed above.

The preferential formation of UmuD-UmuD' heterodimers could have other regulatory consequences. For example, if the heterodimer is indeed inactive or weakly active in SOS mutagenesis, then substantial cleavage of UmuD would have to occur before the active UmuD'$_2$ homodimer would be produced in quantity. This would represent an additional level of regulation of the appearance of the activity necessary for mutagenesis beyond transcriptional derepression of the *umuDC* operon and posttranslational activation of UmuD by RecA-mediated cleavage. It is also possible that formation of UmuD-UmuD' heterodimers could influence RecA-mediated cleavage of UmuD either positively or negatively.

It has been postulated that UmuD' and UmuC proteins function by directly modifying a DNA polymerase in a fusion that permits that polymerase to bypass DNA adducts that would otherwise constitute a block to DNA replication (9, 20, 24, 26). Given the elaborate control circuitry that has evolved to regulate the appearance of this activity seems reasonable that it might also have mechanisms for eliminating the potentially mutagenic effect caused by such a modified polymerase after its function is complete. Deactivating UmuD' by heterodimer formation with intact UmuD may represent such a mechanism.

groE MUTANTS OF E. coli ARE DEFECTIVE IN umuDC-DEPENDENT MUTAGENESIS

Recently we discovered a requirement for the GroE proteins for UV and chemical mutagenesis in E. coli (13). We had made the observation that overexpression of the umuDC operon causes E. coli cells to become cold-sensitive for growth (35). Cells that overproduce UmuD and UmuC proteins as a consequence of the presence of both a lexA51(Def) mutation and a multicopy plasmid carrying the umuD⁺C⁺ operon grow well at 42°C. However, when the temperature is shifted to 30°C, DNA synthesis stops immediately and ultimately the cells die. We were able to produce this phenomenon by expressing the umuDC operon by the IPTC-inducible promoter, P_{trc}. We found that mutations in several heat-shock genes (lon-146::Tn10, rpoH165, dnaK756, groEL100, and groES30) were able to suppress this cold-sensitivity and that the suppression appeared to occur by at least two different mechanisms. lon and dnaK mutations only suppress cold-sensitivity efficiently in a lexA51(Def) background but not in a lexA71::Tn5 background; this class of suppression appears to occur by stabilization of the proteolytically stable LexA51 protein. In contrast, mutations in groES, groEL and rpoH suppressed cold-sensitivity regardless of the transcriptional regulation of the umuDC genes. These observations led to the discovery that mutants defective in groES and groEL are defective in umuDC -dependent UV mutagenesis and that the defect can be partially suppressed by increased expression of the umuDC operon (13). The mechanism by which groESL mutations affect UmuDC functions may be related to the stability of the UmuC protein, since the half-life of this protein is shortened because of mutations at the groESL locus. Since GroEL and GroES function as molecular chaperones that mediate protein folding and protein-protein interactions, a reasonable model for the role of GroEL and GroES in UV mutagenesis is that they allow the UmuC protein to correctly associate into a complex with other proteins involved in UV mutagenesis.

POSSIBLE BIOCHEMICAL MECHANISMS OF UV AND CHEMICAL MUTAGENESIS

The concept that an altered polymerase/replication apparatus might be involved in UV mutagenesis was proposed by Witkin in 1969 (54). This hypothesis was later expanded by Radman's suggestion that SOS-regulated proteins are involved in this process (42). Extensive analyses of DNA sequence changes resulting from UV or chemical mutagenesis (37,51) have indicated that umuD⁺C⁺-dependent mutagenesis is targeted and have therefore supported the concept that a key event in such mutagenesis is the misincorporation of bases opposite noncoding or potentially noncoding lesions. On the basis of a set of physiological experiments involving photoreversal of pyrimidine dimers, Bridges and Woodgate (8, 9) proposed a two-step model for mutagenesis: a umuC⁺-independent step in which an incorrect nucleotide (or nucleotides) is inserted opposite a premutagenic lesion and subsequent umuC⁺-dependent step in which chain elongation is continued after the misincorporated nucleotide.

The identity of the polymerase (or polymerases) involved in this putative mechanism for mutagenesis is still unclear. DNA polymerase I is not required since Bates et al. (2) have recently shown that polA strains that completely lack DNA polymerase I are mutable by UV. Apparently the error-prone from of DNA polymerase I present in SOS-induced cells (29) is not necessary for mutagenesis. Genetic and physiological evidence consistent with the view that DNA polymerase III is required for UV mutagenesis has been presented by Bridges and Mottershead (7) and Hagensee et al. (23). There is a possibility that DNA polymerase II is involved, especially since Bonner *et al.* (6) have reported the purification from SOS-induced cells of an 84 kDa DNA polymerase activity, which they believe to be DNA polymerase II, that is capable of insertion and bypass at apurinic sites. Until this issue is resolved, it is also worth remembering the suggestion of Cupido (11) that translesion synthesis could in principle be carried out by an RNA polymerase rather than a DNA polymerase.

POSSIBLE BIOCHEMICAL ROLES FOR UmuD AND UmuC IN MUTAGENESIS

As mentioned above, there have been a number of suggestions for possible roles of UmuD and UmuC in UV mutagenesis including i) inhibiting proofreading by complexing with the epsilon subunit of DNA polymerase III and aiding misincorporation opposite a lesion (20, 26), ii) helping a polymerase/replication complex continue replication once a base has been misincorporated (8, 24), or iii) helping a polymerase to reinitiate once it has dissociated at a lesion (32). Our observation that overexpression of the *umuDC* operon inhibits DNA replication at 30°C is consistent with an interaction of these proteins with a polymerase or replication complex. To date, there is no direct evidence for any of UmuD or UmuC carrying out any of these roles. Nor, despite many years of effort by several labs, has anyone yet succeeded in establishing an *in vivo* system for UV or chemical mutagenesis that is *umuD$^+$C$^+$*-dependent. We have observed amino acid similarities between UmuD and UmuC and the DNA polymerase accessory proteins, gp45, gp44, and gp62, of bacteriophage T4 (3, 4). The complex of the gp45/gp44/gp62 proteins is known both to influence the 3'-5'-proofreading exonuclease of T4 polymerase and to function as a "sliding clamp" that makes the polymerase much more processive (34). The properties of these bacteriophage proteins thus suggest roles for UmuD and UmuC in mutagenesis that are consistent with the first two of these suggestions.

ACKNOWLEDGEMENTS

We thank the members of our research group for many helpful discussions. This work was supported by Public Health Service Grant CA21615 awarded by the National Center Institute and GM28988 awarded by the National Institute of General Medical Sciences. J.R.B. was supported in part by postdoctoral fellowships from the American Cancer Society, Massachusetts Division and the National Institutes of Health. C.E.D. was supported by postdoctoral fellowship PF3017 awarded by the American Cancer Society.

LITERATURE

1. Bagg, A., Kenyon, C. J., & Walker, G.C. (1981) *Proc. Natl. Acad. Sci. USA 78*, 5749-5753.
2. Bates, H., Randall, S.K., Rayssiguier, C., Bridges, B. A., Goodman, M. F. and Radman, M. (1989) *J. Bacteriol. 171*, 2480-2484.
3. Battista, J.R., Nohmi, T., Donnelly, C.E. and Walker, G.C. Role of UmuD and UmuC in UV and Chemical Mutagenesis. *In:* "Mechanisms and Consequences of DNA Damage Processing", (Eds.) E.C. Friedberg, P.C. Hanawalt, pp 455-459, Alan R. Liss, Inc., New York, (1988a).
4. Battista, J.R., Nohmi, T., Donnelly, C.E., and Walker, G.C. (1989) Amino Acid Similarities to Other Proteins Offer Insights Into Roles of UmuD and UmuC In Mutagenesis. *Genome 31*, 594-596.
5. Battista, J. R., Ohta, T., Nohmi, T., Sun, W., and Walker, G.C. (1990) *Proc. Natl. Acad, Sci. USA 87*, 7190-7194.

6. Bonner, C. A., Randall, S. K., Rayssiguier, C., Radman, M., Eritja, R., Kaplan, B.E., McEntee, K. and Goodman, M.R. (1988) *J. Biol. Chem. 263*, 18946-18952.

7. Bridges, B.A., & Mottershead, R.P. (1978) *Mol. Gen. Genet. 162*, 35-41.

8. Bridges, B. A. & Woodgate, R. (1984) *Mol. Gen. Genet. 196* 364-366.

9. Bridges, B. A. & Woodgate, R. (1985) *Proc. Natl. Acad. Sci USA 82*, 4193-4197.

10. Burckhardt, S. E., Woodgate, R., Scheuermann, R. H., & Echols, H. (1988) *Proc. Natl. Acad. Sci. USA 85*, 1811-1815.

11. Cupido, M. (1983) *Mutat. Res. 109*, 1-11.

12. Defais, M., Caillet-Fauquet, P., Fox, M. S., & Radman, M. (1976) *Mol. Gen. Genet. 148*, 125-130.

13. Donnelly, C., and Walker, G.C. (1989) *J. Bacteriol. 171*, 6117-6125.

14. Dutreix, M., Moreau, P. L., Bailone, A., Galibert, F., Battista, J. R., Walker, G. C. and Devoret, R. (1989) *J. Bacteriol. 171*, 2415-2423.

15. Eguchi, Y., Ogawa, T., & Ogawa, H. (1988) *J. Molec. Biol. 202*, 565-574.

16. Elledge, S.J. & Walker, G.C. (1983) *J. Molec. Biol. 164*, 175-192.

17. Elledge, S.J. & Walker, G.C. (1983) *J. Bacteriol. 155*, 1306-1315.

18. Ennis, D.G., Ossanna, N., and Mount, D.W. (1989) *J. Bacteriol. 171*, 2533-2541.

19. Fersht, A. R. & Knill-Jones, J.W. (1983) *J. Molec. Biol. 165*, 669-682.

20. Foster, P.L., Sullivan, A.D., & Franklin, S. B. (1989) *J. Bacteriol. 171*, 3144-3151.

21. Gimble, F.S., and Sauer, R.T. (1986) *J. Molec. Biol. 192*, 39-47.

22. Gimble, F.S., and Sauer, R.T. (1989) *J. Molec. Biol. 206*, 29-39.

23. Hagensee, M.E., Timme, T., Bryan, S., & Moses, R. (1987) *Proc. Natl. Acad. Sci. USA 84*, 4149-4199.

24. Hevroni, D. and Livneh, Z. (1988) *J. Biol. Chem. 85*, 5046-5050.

25. Hunkapiller, M. W., Smallcombe, S. H., Whitaker, D. R., and Richards, J. H. (1973) *Biochemistry 12*, 4732.

26. Jonczyk, P., Fijalkowska, I., & Ciesla, Z. (1988) *Proc. Natl. Acad. Sci. USA 85*, 9124-9127.

27. Kato, T. & Shinoura, Y. (1977) *Mol.Gen.Genet. 156*, 121-131.

28. Kitagawa, Y., Akaboshi, E., Shinagawa, H., Horii, T., Ogawa, H. & Kato, T. (1985) *Proc. Natl. Acad. Sci. USA 82*, 4336-4340.

29. Lackey, D., Krauss, S.W., & Linn, S. (1982) *Proc. Natl. Acad Sci. USA 79*, 330-334.

30. Lin, L.-L. & Little, J.W. (1988) *J. Bacteriol. 170*, 2163-2173.

31. Little, J.W. (1984) *Proc. Natl. Acad. Sci. USA 81*, 1375-1379.

32. Livneh, Z. (1986) *J. Biol. Chem. 261*, 9526-9533.

33. Lu, C., Scheuermann, H. & Echols, H. (1986) *Proc. Natl. Acad. Sci. USA 83*, 619-623.

34. Mace, D. C. & Alberts, B. M. (1984) *J. Molec. Biol. 177*, 279-293.

35. Marsh, L. & Walker, G.C. (1985) *J. Bacteriol. 162*, 155-161.

36. McCann, J., Choi, E., Yamasaki, E., Ames, B.N. (1975) *Proc. Natl. Acad,. Sci. USA 72*, 5135-5139.

37. Miller, (1983) *Ann. Rev. Genet. 12*, 215-238.

38. Molina, A. M., Babulri, N., Tamaro, M. Venturini, S. & Monti-Bragadin, C. (1979) *FEMS Microbiol. Lett. 5*, 33-37.

39. Nohmi, T., Battista, J. R., Dodson, L. A. and Walker, G.C. (1988) *Proc. Natl. Acad. Sci. USA 82*, 4331-4335.

40. Perry, K. L., Elledge, S. J., Mitchell, B. B., Marsh, L. & Walker, G. C. (1985) *Proc. Natl. Acad. Sci. 82*, 4331-4335.

41. Perry, K. L. & Walker, G. C. (1982) *Nature (London) 300*, 278-281.

42. Radman, M. (1974) In *Molec. and Environ. Aspects of Mutagen.*, edited by L. Prakash,

F. Sherman, M. Miller, C. Lawrence, and H. W. Tabor (Springfield, IL., Charles C. Thomas, Pub), 128-142.

43. Sassanfar, M., & Roberts, J. W. (1990) *J. Molec. Biol. 212*, 79-96.

44. Sauer, R. T., Yocum, R. R., Doolittle, R. F., Lewis, M. & Pabo, C. O. (1982) *Nature 298*, 447-451.

45. Sedgwick, S. G. & Goodwin, P. A. (1985) *Proc. Natl. Acad. Sci USA 82*, 4172-4176.

46. Shinagawa, H., Iwasaki, H., Kato, T., & Nakata, A. (1988) *Proc. Natl. Acad. Sci. USA 85*, 1806-1810.

47. Slilaty, S. N. & Little, J. W. (1987) *Proc. Natl. Acad. Sci. USA 84*, 3987-3991.

48. Steinborn, G. (1978) *Mol. Gen. Genet. 165*, 87-93.

49. Strike, P. & Lodwick, D. (1988) *J. Cell. Biochem. (Suppl. 12A*, 326).

50. Strike, P., & Lodwick, D. (1987) *J. Cell. Sci, (Suppl. 6*, 303-321).

51. Walker, G. C. (1984) *Microbiol. Rev. 48*, 60-93.

52. Walker, G.C. (1985) *Ann. Rev. Biochem. 54*, 425-457.

53. Weigle, J. J. (1953) *Proc. Natl. Acad. Sci. USA 39*, 628-636.

54. Witkin, E. M. (1969) *Ann. Rev. Microbiol. 23*, 487-514.

55. Witkin, E. M. (1975) *Mol. Gen. Genet. 142*, 87-103.

INDUCTION OF CAT mRNA TRANSLATION BY CHLORAMPHENICOL: AN EXAMPLE OF TRANSLATIONAL ATTENUATION

Paul S. Lovett, Nicholas P. Ambulos, Jr., and Elizabeth J. Rogers

Department of Biological Sciences, University of Maryland
Catonsville, MD 21228

INTRODUCTION

Virtually all gene regulatory schemes fall into one of two categories; those that regulate at the transcriptional level and those which regulate translation. Most gene regulatory devices that are known modulate transcription, although in recent years an increasing number of translational control systems have been identified (5). The elucidation of a wide range of gene control systems has typically clarified our understanding of those fundamental processes that are needed to phenotypically express a gene. This is most clearly seen in the case of transcription, but also applies to translation control. cat gene regulation by an antibiotic is one example of a translational control device termed translational attenuation (7). Since the regulatory device depends on antibiotic perturbation of translation, solution of the control mechanism may reveal aspects of antibiotic action and ribosome function that cannot be approached by conventional methodology.

REGULATION OF CAT-86, A GENE THAT SPECIFIES CHLORAMPHENICOL ACETYLTRANSFERASE

Chloramphenicol acetyltransferase (CAT) is an intracellular enzyme that acetylates chloramphenicol and thereby inactivates the antibiotic action of the drug (11). Genes that encode CAT are designated cat, and cat genes from gram-positive bacteria are typically induced by chloramphenicol. cat-86 is an example of a regulated, gram-positive cat gene that was cloned onto a high copy plasmid in Bacillus subtilis (7). Gene fusion and deletion studies showed that the regulated phenotype associated with cat-86 resides within an 84bp segment located immediately 5' to the cat-86 coding sequence. This regulatory region consists of two domains. This first is a stable stem-loop structure that is predicted to sequester the cat-86 ribosome binding site in mRNA. Second, the stem-loop is preceded by a short open-reading frame, the leader, with its own ribosome binding site and initiation codon. Deletions whose end points are within the leader prevent cat-86 expression, while deletions that extend into the left leg of the stem-loop structure cause high-level constitutive cat expression (7). Consequently the stem-loop must block cat-86 expression and the regulatory leader must enable chloramphenicol to overcome the block.

Biotechnology and Environmental Science: Molecular Approaches
Edited by S. Mongkolsuk et al., Plenum Press, New York, 1992

The RNA synthesis inhibitors rifampin and streptolydigin block all detectable RNA synthesis in B. subtilis within 5 minutes. Yet, cat-86 induction by chloramphenicol can be demonstrated for more than 30 minutes after inhibition of RNA synthesis (3). Thus, induction of cat-86 results from a form of regulation that controls translation.

The cellular target for cat-86 inducers, i.e., chloramphenicol, amicetin and erythromycin, is the 50S subunit of bacterial ribosomes (4,10,11). Mutations in the B. subtilis chromosome that prevent the inducer from binding to the 50S subunit or a mutation that deletes a protein subunit from the 50S subunit diminish or abolish induction (2,4). Thus induction appears to require ribosome participation.

The transcriptional attenuation model proposes a scheme in which ribosome stalling in a regulatory leader alters the secondary structure of a downstream transcription terminator, allowing read through transcription (6). To test the notion that ribosome stalling in the cat-86 leader may be the mechanism that leads to induction, ribosomes were stalled at discrete leader codons by amino acid starvation (1). This methodology revealed that when the aminoacyl (A) site of the ribosome was stalled at leader codon 6, cat-86 induction was observed, without the addition of an exogenous inducer. Similarly, when leader codon 6 was mutated to a termination codon UAA, cat-86 was inducible by chloramphenicol, but no induction was possible when earlier leader codons were replaced with the ochre codon (1). Hence induction is due to ribosome stalling in the regulatory leader and the precise site of stalling was established by the amino acid starvation experiments; the A site must occupy leader codon 6.

THE STALL SEQUENCE IS SPECIFIC AND COMPLEMENTARY TO 16S rRNA

Chloramphenicol might induce cat-86 by stalling a ribosome at a specific leader sequence that places the A site at leader codon 6. Alternatively, the antibiotic might randomly stall ribosomes at various sites within the leader but induction would result only from those ribosomes stalled with their A sites at codon 6. To distinguish between these models a second copy of leader codon 5, GAT, was inserted between leader codon 5 and 6. This 5A codon insertion prevented drug induction, and induction was subsequently restored by deleting leader codon 6. These data eliminate the random stalling model and support sequence specificity in the site of stalling. To identify the stall sequence the leader sequence 5' to codon 6 was resynthesized stepwise between leader codons 5 and 6 (9). By this approach leader codons 2 through 5 were found to constitute the stall sequence which we now refer to as the crb-86 box.

The crb-86 box may stall drug-sensitized ribosomes by encoding a leader peptide whose amino acid sequence contributes to ribosome stalling. This predicts that the nature of the amino acids in the peptide may be essential to stalling. Consistent with this proposal is the observation that missense mutations in the leader that result in substituting amino acids with non-like amino acids diminish induction. However, we have also considered a potential additional role for crb. crb-86 might guide the ribosome to the precise site of stalling by an interaction between the nucleotide sequence of crb-86 with a sequence in rRNA. Inspection of the sequence of 16S rRNA of B.subtilis reveals a complement to crb-86, and not surprisingly this 1 6S rRNA sequence also contains the complement to leader sequences for erm and tet genes which are, or are likely, regulated by a mechanism similar to cat-86 (8).

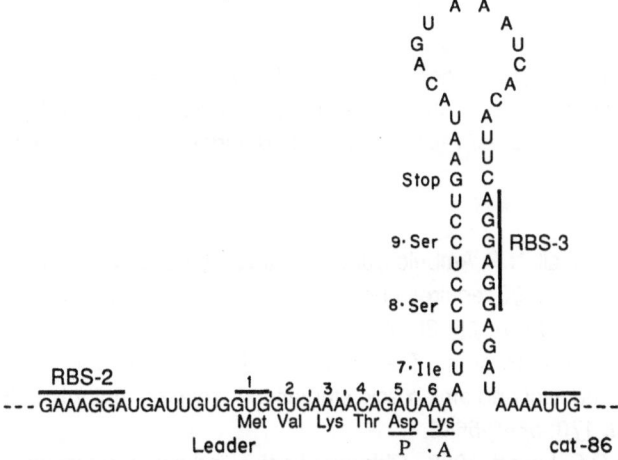

Figure 1. Regulatory region of cat-86 transcripts which is responsible for the chloramphenicol-inducibility. RBS-2 and -3 are the ribosome binding sites for the leader and the cat-86 coding sequence, respectively. P and A are the codons occupied by the peptidyl and aminoacyl sites of a chloramphenicol-stalled ribosome. The codon UUG is the start of the cat-86 coding sequence.

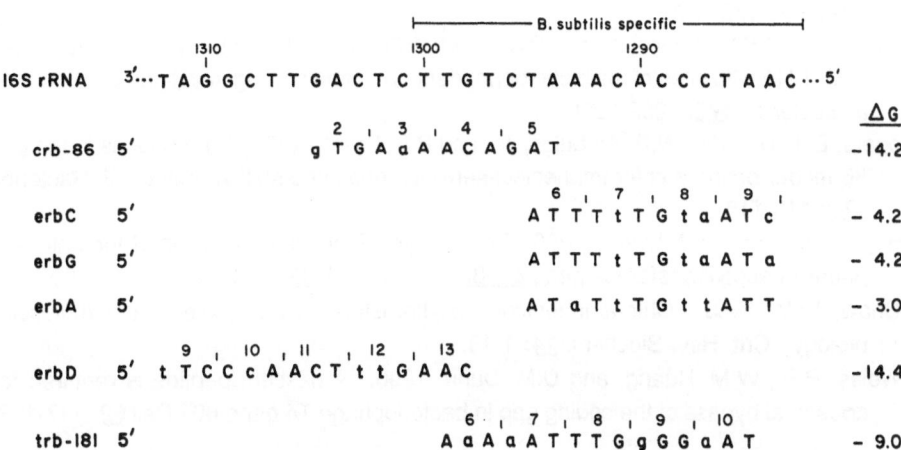

Figure 2. Complementary between 16S rRNA of Bacillus subtilis and sequences in the leaders for cat, erm and tet genes.

CONCLUSIONS

The role of crb-86 in cat-86 induction may be a function of both its complementarity with 16S rRNA and the crb-encoded peptide. Solving the role of crb as a gene regulatory device will dissect ribosome function in a manner not possible by conventional methodology much in the same vein as other recently discovered variations of ribosome function (12).

REFERENCES

1. Alexieva, Z., E.J. Duvall, N.P. Ambulos, Jr., U.J. Kim and P.S. Lovett. 1988. Chloramphenicol induction of cat-86 requires ribosome stalling at a specific site in the leader Proc. Natl. Acad. Sci. U.S.A. 85: 3057-3061.
2. Ambulos, N.P., Jr., E.J. Rogers, Z. Alexieva, and P.S. Lovett 1988. Induction of cat-86 by chloramphenicol and amino acid starvation in relaxed mutants of Bacillus subtilis. J. Bacteriol. 170: 5642-5646.
3. Duvall, E.J. and P.S. Lovett. 1988. Chloramphenicol induces translation of the mRNA for a chloramphenicol resistance gene in Bacillus subtilis. Proc. Natl. Acad. Sci. U.S.A. 83: 3939-3943.
4. Duvall, E.J., S. Mongkolsuk, U.J. Kim, P.S. Lovett, T.M. Henkin, and G.H. Chambliss. 1985. Induction of the chloramphenicol acetyltransferase gene cat-86 through the action of the ribosomal antibiotic amicetin. Involvement of a Bacillus subtilis ribosomal component in cat induction. J. Bacteriol. 161: 665-672.
5. Gold, L. 1988. Post-transcriptional regulatory mechanisms in Escherichia coli. Annu. Rev. Biochem. 57: 199-233.
6. Landick, R., and C. Yanofsky. 1987. Transcription attenuation, p. 1276-1301. In F.C. Neidhardt, J.L. Ingraham, B. Magasanik, K.B. Low, M. Schaechter, and H.E. Umbarger (Eds), Escherichia coli and Salmonella typhimurium: Cellular and Molecular Biology. American Society for Microbiology, Washington, D.C.
7. Lovett, P.S. 1990. Translational attenuation as the regulator of inducible cat genes. J. Bacteriol. 172: 1-6.
8. Rogers, E.J., N.P. Ambulos, Jr., and P.S. Lovett. 1990. Complementarity of Bacillus subtilis 16S rRNA with sites of antibiotic-dependent ribosome stalling in cat and erm leaders. J. Bacteriol. 172: 6282-6290.
9. Rogers, E.J., U.J. Kim, N.P. Ambulos, Jr., and P.S. Lovett. 1990. Four codons in the cat-86 leader define a chloramphenicol-sensitive ribosome stall sequence. J. Bacteriol. 172: 110-115.
10. Rogers, E.J. and P.S. Lovett. 1990. Erythromycin induces expression of the chloramphenicol acetyltransferase gene cat-86. J. Bacteriol. 172: 4694-4695.
11. Shaw, W.V. 1983. Chloramphenicol acetyltransferase: enzymology and molecular biology. Crit. Rev. Biochem. 14: 1-43.
12. Weiss, R.B., W.M. Huang, and D.M. Dunn. 1990. A nascent peptide is required for ribosomal bypass of the coding gap in bacteriophage T4 gene 60. Cell 62: 117-126.

GENETIC ANALYSIS OF BACTERIAL STRAINS FRESHLY ISOLATED FROM NATURAL SOURCES

B. W. Holloway, A. Bowen, S. Dharmsthiti, V. Krishnapillai,
A. Morgan, E. Ratnaningsih and M. I. Sinclair

Department of Genetics and Developmental Biology
Monash University, Clayton, Victoria, Australia

Microbiology is an essential component of new developments in biotechnology and environmental science and, while the genetic knowledge available for *Escherichia coli, Bacillus subtilis* and *Pseudomonas* provides an essential knowledge base for these developments, there is a need for genetic dissection of microorganisms newly isolated from nature. The aim of such genetic work could include an increase in the yield of a valuable gene product, to alter the specificity of an enzyme or to stabilize a particular phenotype over a large number of generations. The application of genetic techniques, particularly the more classical procedures may be ineffectual with newly isolated strains and experience in our own and other laboratories has shown the strain specificity of these approaches.

Over a number of years our laboratory has had a need to manipulate by genetic means a variety of different bacteria. This has led us to test a range of procedures by which genetic information can be obtained and genetic alterations made to selected properties. We have concluded that it is no longer necessary to establish the classical techniques of conjugation, transduction and transformation for the genetic manipulation of new bacteria. Instead the use of physical approaches and the manipulation of cloned DNA in selected surrogate hosts provide new ways of obtaining novel strains. This helps circumvent unforeseen barriers present in newly isolated organisms, enabling a concentration on those parts of the genome of the organism which are of particular interest. Furthermore, by the use of pulsed field gel electrophoresis, whole bacterial genomes can now be examined.

TECHNIQUES FOR GENETIC MODIFICATION

There are specific techniques which when used appropriately can be highly effective in increasing the effectiveness of organisms for biotechnology and for solving environmental problems.

TRANSPOSON MUTAGENESIS

Tn5 is the most useful transposon for gram negative bacteria as many organisms found in nature have low levels of kanamycin resistance. If

Biotechnology and Environmental Science: Molecular Approaches
Edited by S. Mongkolsuk et al., Plenum Press, New York, 1992

169

necessary, double antibiotic selection may be imposed or kanamycin hypersensitive mutants isolated as has been done with *P. cepacia* (H. Matsumoto, personal communication). Appropriate vectors for introducing Tn*5* into a variety of bacteria include *E. coli* narrow host range plasmids such as pACYC184, pACTC177 or pBR325 which include the mobilization functions of the promiscuous IncP-1 plasmid RP4, enabling conjugative transfer to other gram-negative bacteria and Tn*5* mutagenesis (Simon *et al.*, 1986). Other transposon vectors are temperature sensitive plasmid replication defective mutants of the promiscuous IncP-1 plasmids RP1 or RP4 (Rella *et al.*, 1985; O'Hoy and Krishnapillai, 1985, 1987) and the *P. aeruginosa* IncP-10 plasmid pMO75 (Whitta *et al.*, 1985; Strom *et al.*, 1990). The latter is a derepressed transfer mutant of the *P. aeruginosa* IncP-10 plasmid R91-5 carrying Tn*5*. It has a wide host range for transfer but its replication is restricted to some strains of *P. aeruginosa*. Matings of *P. aeruginosa* PAO(pMO75) with a variety of gram negative genera including *P. putida*, *P. syringae*, *P. solanacearum*, *Methylobacterium* and *Flavobacterium* have resulted in mutants although the recovery of mutants varies from organism to organism. The method is not applicable to gram positive genera.

GENOMIC LIBRARIES

The wide host range mobilizable cosmid pLA2917 (Allen and Hanson, 1985) has been found to be suitable for various genera including *Pseudomonas*, *Arthrobacter*, *Methylophilus*, *Methylobacterium* and *Flavobacterium*. There is no evidence of chromosome fragment rearrangements, the banks are comprehensive, and the insert size is ca 25 kb. Individual clones are kept as frozen suspensions in microtitre trays at -70ºC. The *E. coli* strain S17-1 is used as the host, this having the IncP-1 wide range plasmid RP4 integrated into the bacterial chromosome (Simon *et al.*, 1986). Thus cosmids can be conjugatively mobilized from the library to other gram negative bacteria by two factor crosses and by means of a multi-pronged inoculator, thousands of matings can be done in a few hours. The wide host range property of the vector enables complementation tests with a variety of gram negative surrogate hosts to be accomplished.

GENE REPLACEMENT

Gene replacement enables mutated or genetically modified segments of DNA to be inserted into selected bacterial genomic sites. There is reciprocal recombination between homologous sequences shared by an incoming cloned DNA fragment and the genomic target. Systems in which the region of homology can be interrupted by passenger genes are of particular use and transposon insertions generated in cloned DNA within *E. coli* can be used for site-specific mutagenesis within the original host. Novel clusters of foreign genes constructed *in vitro* can be inserted stably and precisely into chosen genomic sites of the host.

There are a range of gene replacement techniques for bacteria, but in practice many are strain specific with a low frequency of reintegration. Using *P. solanacearum*, *Flavobacterium* and *P. putida*, we have shown that recombinant cosmids derived from pLA2917 show a readiness for passenger DNA to be reintegrated into the chromosome at the appropriate homologous region. By appropriate cosmid and donor parent selection strategies up to 50% of transconjugant colonies (10-50%) are found to have undergone gene replacement and lost both the wild-type region and the vector (A. Bowen, R. Qi, D. Maris and D. Strom, unpublished data). These findings provide an additional argument for the use of pLA2917 in constructing cosmid libraries.

COMPLEMENTATION

A gene of interest from a newly isolated bacterium can be identified by its ability to restore a mutant defect or confer a new phenotype in a surrogate host, although precise identification of defective function requires a demonstration that the homologous gene product is missing. There are two ways of doing this. Firstly by restoration of wild type function of the mutant and this can be done readily by complementation of a mutant of the homologous bacterium by a cloned genomic fragment. The vector carrying the fragment must be able to be transferred from the cloning host to the mutant strain. Given that *E. coli* has the best range of vectors, a wide host range cosmid vector such as pLA2917 has many advantages.

The second way is to use a surrogate host for which mutants of the relevant phenotype have been characterized. Not all bacteria are suitable, for example, *E. coli* does not express genes from other genera particularly well while by contrast, pseudomonads express heterologous DNA very efficiently (Jeenes *et al.*, 1986). This property has been used for the complementation mapping of the obligate methylotrophs *Methylophilus methylotrophus* and *M. viscogenes* (Lyon *et al.*, 1986).

If no suitable complementation system is available, or if the phenotype cannot be scored, then the gene being sought can be detected by scanning of gene libraries using Western blots for translationally expressed genes or by hybridization with oligonucleotides derived from amino acid sequence data.

NEW APPROACHES FOR GENETIC MAPPING

Some knowledge of genome arrangement may be necessary for the effective manipulation of bacterial strains. For example, are the genes of interest located on the chromosome or on a plasmid? Are the genes of a metabolic pathway clustered or scattered? How many genes are involved in a particular metabolic pathway given only limited knowledge of the intermediary compounds?

Such information has usually been derived from genetic mapping using gene exchange systems - conjugation, transduction and transformation. The development of recombinant DNA techniques has made possible the physical mapping of cloned fragments of DNA, but until recently genetic exchange techniques were still considered a necessary component of genome analysis. With the advent of pulsed field gel electrophoresis (PFGE), which permits the separation of DNA fragments in the megabase range, restriction maps of whole bacterial chromosomes can be constructed. At their simplest, such maps allow accurate measurement of genome size and can reveal aspects of genome complexity, such as the existence of two chromosomes in *Rhodobacter sphaeroides* (Suwanto and Kaplan, 1989). Whole genome analysis using PFGE has been carried out with a number of genera including *Escherichia*, *Pseudomonas, Anabaena, Caulobacter, Mycoplasma, Haemophilus, Bacillus, Methylobacterium, Streptococcus* and *Clostridium*. Up to about 40 fragments can be resolved by PFGE, the choice of restriction enzymes depending on nucleotide content and the pattern of restriction sites. Enzymes used include *Spe*I, *Not*I, *Sfi*I, *Dra*I, *Ase*I, *Ssp*I, *Avr*II, *Xba*I, *Nhe*I, *Rsr*II and *Dpn*I and the general techniques for mapping of bacterial genomes by this procedure have been reviewed by Smith and Condemine (1990).

We have worked on three species of *Pseudomonas, P. aeruginosa, P. putida* and *P. solanacearum*. For *P. aeruginosa* 37 *Spe*I fragments were detected ranging in size from 525 kb - 5.1 kb giving a genome size of 5862 kb. A combined physical and genetic map was generated by probing with cosmid clones from a genomic library prepared using pLA2917 carrying known chromosomal genes. In some regions Tn1 inserts were used to identify some *Spe*I fragments. Contiguity of *Spe*I fragments was demonstrated by probing with *Spe*I linking clones which carry DNA from either side of a *Spe*I site, or by probing *Spe*I partial digests (E. Ratnaningsih, S. Dharmsthiti, V. Krishnapillai, A. Morgan, M. Sinclair and B. Holloway, 1990; Holloway *et al.*, 1990a).

For *P. putida,* chromosomal transposon inserts have proved to be more useful than cosmid clones due to the ease with which Tn5 inserts can be mapped genetically (Strom *et al.*, 1990). So far 39 *Spe*I fragments have been identified, from 710 kb - 0.6 kb in size, and totalling 5846 kb. Nearly 90% of the genome has been assigned to chromosomal locations and correlated with the genetic map (R. Saffery and A. Morgan, unpublished data).

P. solanacearum has no developed genetic system, and no genetic map. We are constructing a combined physical and genetic map by using PFGE in combination with complementation mapping of mutants of *P. aeruginosa* by genes carried on cosmid clones. 29 *Spe*I fragments have now been identified, from 470-12 kb to give a genome size of 5903 kb (D. Escuadra and B. Holloway, unpublished data).

It is possible to generate a combined physical and genetic map without the availability of any mutants or even a cosmid clone library. A physical map could be constructed using linking clones and/or partial digestion, and then genetic markers placed on the map by probing with cloned genes (or oligonucleotides where the sequence is known) either from the bacterium in question, or from other species. Some types of genes show sufficient sequence conservation to allow a positive hybridization signal. In *P. aeruginosa* we have mapped a number of genes in this way including those for RNA polymerase, the origin of replication, ribosomal RNA and threonine biosynthesis (Holloway *et al.*, 1990b).

PLASMID-CHROMOSOME RELATIONSHIPS

Plasmids are frequently a component of the bacterial genome and genes of interest for biotechnology or environmental microbiology may be located on a plasmid or on the chromosome. Isolation of plasmid DNA can be difficult in newly isolated organisms, particularly if a large (>200 kb) plasmid is involved. PFGE has been used to demonstrate plasmids of varying sizes in the *Rhodobacter sphaeroides* genome (Suwanto and Kaplan, 1989).

The TOL plasmid (pWWO) has been extensively studied in a variety of forms from different strains of *Pseudomonas* and is a useful model for the techniques by which degradative plasmids can be genetically manipulated to result in bacterial strains with modified metabolic activities.

A strain of *P. putida* MW1000 has been used by Celgene Corporation for the commercial production of cis, cis-muconic acid (Maxwell, 1986). The *P. putida* strain MW1000 used in the production of cis, cis-muconic acid has the 56 kb transposon Tn*4651* (Tsuda and Iino, 1988) from the TOL plasmid integrated into the bacterial chromosome. The site of insertion has been mapped to a region of the *P. putida* chromosome containing the genes from the β-ketoadipate pathway, which shares functional similarity with part of the TOL catabolic pathway (Sinclair *et al.*, 1986). There is substantial DNA sequence homology between the TOL plasmid *xylXYZ* gene cluster and the chromosomal *ben* genes of *P. putida* and *P. aeruginosa* (M.I. Sinclair and B. Holloway, unpublished data).

A simple temperature selection protocol has been developed to allow the isolation of TOL DNA insertions into the chromosome of the genetically well characterized strain *P. aeruginosa* PAO. Analysis of these insert strains by PFGE has revealed a marked regional specificity of insertion, with all inserts so far examined being confined to a region of approximately 300 kb, representing less than 6% of the PAO chromosome.

The combined PFGE and probing techniques used to map these TOL inserts has provided more accurate locations than could have been obtained by conventional techniques such as Hfr donors and phenotypic identification of recombinants was less time consuming.

CONCLUSIONS

The increased availability and a variety of recombinant DNA techniques, has provided a route for the genetic analysis of newly isolated bacteria. The data so obtained is essential if the manipulation of metabolic pathways of these strains is to be successful.

Three general conclusions are possible. Using genetically well characterized models, effective procedures for the genetic analysis of newly isolated bacteria can be developed. Next, a combination of physical and genetic techniques provides genetic analysis superior to the more traditional procedures. Finally, improved techniques enables whole genome analysis to be applied to a range of organisms. The unveiling of hidden complexities in the manner in which bacteria store their biological information can only contribute to our ability to manipulate them for practical purposes.

ACKNOWLEDGEMENTS

Work in the authors' laboratory is supported by the Australian Research Council, Celgene Corporation and Imperial Chemical Industries (Australia).

REFERENCES

Allen, L.N. and Hanson, R.S. (1985). J. Bacteriol. 161, 955-962.

Holloway, B.W., Dharmsthiti, S., Johnson, C., Kearney, A., Krishnapillai, V., Morgan, A.F., Ratnaningsih, E., Saffery, R., Sinclair, M., Strom, D. and Zhang, C. (1990a). In *Pseudomonas*: Biotransformations, Pathogenesis and Evolving Biotechnology, pp. 269-279. Ed. S. Silver, A. Chakrabarty, B. Iglewski and S. Kaplan. Washington, D.C. American Society for Microbiology.

Holloway, B.W., Bowen, A., Dharmsthiti, S., Krishnapillai, V., Morgan, A., Ratnaningsih, E. and Sinclair, M. (1990b). Genetic tools for the manipulation of metabolic pathways. Proc. 6th Int. Symp. Genet. Indust. Microorganisms, Strasbourg 227-238.

Jeenes, D.J., Soldata, L., Baur, H., Watson, J.M., Mercenier, A., Reimmann, C., Leisinger, T. and Haas, D. (1986). Mol. Gen. Genet. 203, 421-429.

Lyon, B.R., Kearney, P.P., Sinclair, M.I. and Holloway, B.W. (1988). J. Gen. Micro. 134, 123-13

Maxwell, P.C. (1986). US Patent 4,588,688.2.

O'Hoy, K. and Krishnapillai, V. (1985). FEMS Micro. Lett. 29, 299-303.

O'Hoy, K. and Krishnapillai, V. (1987). Genetics 115, 611-618.

Ratnaningsih, E., Dharmsthiti, S., Krishnapillai, V., Morgan, A., Sinclair, M. and Holloway, B.W. (1990). J. Gen. Micro. 136, 2351-2357.

Simon, R., Preiefer, U. and Pühler, A. (1983). Bio/Technology 1, 784-791.

Simon, R., O'Connell, M., Labes, M., and Pühler, A. (1986). Meth. Enzymol. 118, 640-659.

Simon, R., Quandt, J. and Klipp, W. (1989). Gene 80, 161-169.

Sinclair, M.I., Maxwell, P.C., Lyon, B.R. and Holloway, B.W. (1986). J. Bacteriol. 168, 1302-1308.

Smith, C.L. and Condemine, G. (1990). J. Bacteriol. 172, 1167-1172.

Strom, D.A., Hirst, R., Petering, J. and Morgan, A. (1990). Genetics 126, 497-503.

Suwanto, A. and Kaplan, S. (1989). J. Bacteriol. 171, 5850-5859.

Tsuda, M. and Iino, T. (1987). Mol. Gen. Genet. 210, 270-276.

Tsuda, M. and Iino, T. (1988). Mol. Gen. Genet. 213, 72-77.

Whitta, S., Sinclair, M.I. and Holloway, B.W. (1985). J. Gen. Micro. 131, 1547-1549.

TRANSLATION INITIATION FROM
A CUG CODON IN *Bacillus subtilis*

Nicholas P. Ambulos, Jr. and Paul S. Lovett

Department of Biological Sciences, University of Maryland
Catonsville, Maryland 21228 USA

INTRODUCTION

In recent years many gene regulatory mechanisms have been identified which exert control at the level of translation initiation (Lovett, 1990; Gold, 1988; Dubnau, 1984). Sequences in mRNA that regulate translation initiation efficiency include the ribosome binding site and translation initiation codon. The ribosome binding site is complementary to sequences at the 3' end of 16S rRNA and functions to bind 30S ribosomal subunits onto mRNA (Shine and Dalgarno, 1974). The extent of complementarity between the ribosome binding site and 16S rRNA determines in part the efficiency of translation initiation. Mutations in the ribosome binding site which prevent translation initiation can be overcome by introducing compensating mutations in 16S rRNA (Hui and De Boer, 1987, Jacob et al., 1987).

Translation initiation efficiencies are also dependent upon the translation initiation codon and its spatial relation to the ribosome binding site. The ideal translation initiation codon is AUG spaced 8 nucleotides downstream from the ribosome binding site (Gold and Stormo, 1987; Kozak, 1983). Codons GUG and UUG will allow translation to initiate but less efficiently than AUG (Gold and Stormo, 1987; Reddy et al., 1985; Clark and Marcker, 1966). Increases

Table I. Effects of translation initiation codon mutations on expression of <u>cat-86C2</u> in <u>Bacillus subtilis</u> strain BR151.

<u>Initiator codon</u>	<u>CAT sp. act</u>[a]	<u>Efficiency</u>[b]
AUG	20.5	100%
UUG	15.1	74%
GUG	13.3	65%
CUG	6.2	30%
AAA	0.103	0.5%
AAG	0.056	0.3%

[a]Values are CAT specific activities (micromoles per min per mg of protein at 25°C). CAT assays were done by colormetric procedure of Shaw (1975). Protein was measured as described by Bradford (1976).

[b]Efficiencies are expressed as a ratio of CAT activity relative to the value obtained with AUG.

Biotechnology and Environmental Science: Molecular Approaches
Edited by S. Mongkolsuk et al., Plenum Press, New York, 1992

175

or decreases in the spacing between the translation initiation codon and ribosome binding site can diminish the efficiency of translation initiation.

In this study we report conditions which allow CUG to be used as a translation initiation codon in *Bacillus subtilis* with moderate efficiency.

RESULTS AND DISCUSSION

The plasmid encoded cat-86 gene specifying chloramphenicol acetyltransferase, was used as a reporter gene in these studies (Lovett, 1990). CAT specific activity was used as a measure of translational efficiency. Translation of cat-86 initiates with UUG rather than AUG (Duvall et al., 1987). The UUG translation initiation codon is spaced 8 nucleotides downstream from the ribosome binding site, 5'-AGGAGG, which is a perfect complement to a hexanucleotide at the 3' end of *B. subtilis* 16S rRNA. The calculated free energy of the pairing is -16.6 kcal/mol (Tinoco et al., 1973). Wild type cat-86 is inducible by chloramphenicol due to a regulatory mechanism termed translational attenuation (Lovett, 1990). A constitutive version of the gene, cat-86C2, was used in these studies (Laredo et al., 1988).

Table II. Effects of mutations in initiator codons and RBS on __xylE__ expression in **Bacillus subtilis** strain BR151.

__xylE gene__	__Initiator Codon__	__(Sp. act.)__[a]	__Efficiency__[b]
xylE	AUG	3.79	100%
xylE	CUG	0.059	1.6%
xylE2	AUG	6.56	100%
xylE2	CUG	0.784	12%

[a]Values are catechol-2,3-dioxygenase (C23D) specific activities (moles per min per mg of protein at 30°C). C23D assays were done as described by Ray et al., (1985). Protein was measured as described by Bradford (1976).

[b]Efficiencies are expressed as a ratio of C23D activity relative to the value obtained with AUG.

<u>cat-86</u>	5'	... A G G A G G A G A T A A A A T T G ...	△G = -16.6 kcal/mol
<u>xylE</u>	5'	... A A G A G G T G A C G T C A T G ...	△G = -9.4 kcal/mol
<u>xylE2</u>	5'	... A G G A G G A T G A C G T C A T G ...	△G = -16.6 kcal/mol

Figure 1. Sequences of the translation initiation regions for *cat-86, xylE* and *xylE2* (Duvall et al., 1983; Zukowski et al., 1985). The underlined sequences are the putative ribosome sites and translation initiation codons. The G values given are for the ribosome binding site sequences (Tinoco et al., 1973).

Expression of *cat-86* increased by 25% when the translation initiation codon was changed from UUG to AUG by oligonucleotide directed mutagenesis (Table I). Changing the initiation codon to GUG slightly decreased CAT activity. CAT activity was abolished when the initiator codon was changed to either AAA or AAG. Unexpectedly, the codon CUG was found to promote moderately efficient expression of *cat-86*. The level of CAT activity when CUG was the initiator codon was 30% compared to levels observed with AUG. To determine if CUG was indeed the initiator codon, corresponding CAT-86 protein was purified and 13 N-terminal amino acids were determined and found to be the same as previously reported for CAT-86 protein specified by wild type *cat-86* (Laredo et al., 1988). Thus CUG was the initiation codon and specified methionine.

It is conceivable that CUG initiates translation of cat-86 due to some unique feature of the gene. To test this hypothesis xylE, which specifies catechol 2,3-dioxygenase, was mutagenized to change the AUG initiation codon to CUG. This change decreased expression by more than 98% (Table II). The decrease in activity was significantly greater than the decrease observed when the AUG initiation codon of cat-86 was changed to CUG. The ribosome binding site for xylE is 5'-AAGAGG and is spaced 7 nucleotides upstream from the initiation codon. The free energy of pairing with B. subtilis 16S rRNA is -9.4 kcal/mol, which is significantly lower than the strength of pairing between the cat-86 ribosome binding site and 16S rRNA (Figure 1). An attempt to increase efficiency of xylE translation initiation from CUG was made. The ribosome binding site and its spatial relationship to the initiation codon was made to duplicate the initiation region of cat-86 (Figure 1). This mutant version of xylE, called xylE2, exhibited a higher level of expression than xylE regardless of whether the initiation codon was AUG or CUG. However, the increase was significantly greater when CUG was the initiator (Table II). CUG will function with moderate efficiency as a translation initiation codon in B. subtilis when provided with am optimal ribosome binding site properly spaced from the initiator codon.

To determine how well CUG would function in E. coli the 6 initiation codon variants of cat-86 were introduced into E. coli on plasmid pBR322 (Bolivar et al., 1977). As seen in table III, the hierarchy and efficiencies of translation initiation were similar to those previously reported (Reddy et al., 1985). CUG functioned poorly as a translation initiation codon, expressing 3% of CAT levels observed when AUG was the initiation codon. AAA and AAG did not function to initiate translation. Thus CUG will initiate translation of cat-86 in B. subtilis, but not in E. coli.

It has been previously shown that mRNA derived from B. subtilis is translated by E. coli, but mRNA derived from E.coli is only poorly translated by B. subtilis (Sharrock et al., 1979). This difference is presumed to be due to the nature of the ribosome binding site and the initiation of ribosomes on mRNA (McLaughlin et al., 1981). E. coli ribosomes will initiate at ribosome binding sites which are relatively weak, while B. subtilis ribosomes require a strong ribosome binding site. The requirement for a strong ribosome binding site in B. subtilis may reflect differences in the translational machinery. Roberts and Rabinowitz (1989) have reported that B. subtilis contains no analogue of E. coli ribosomal protein S1. E. coli ribosomes which lack S1 require a stronger ribosome binding site to initiate translation. It is possible that in the absence of an S1-like protein, translation initiation in B. subtilis requires more extensive mRNA-rRNA interaction in the region of the ribosome binding site. This extensive interaction may result in more efficient utilization of non-AUG translation initiation codons.

Table III. Effects of initiator codon mutations on expression of **cat-86C2** in **E. coli** strain JM105.

Initiator Codon	CAT sp. act.[a]	Effciency[b]
AUG	1.21	100%
GUG	0.59	49%
UUG	0.46	38%
CUG	0.038	3%
AAA	0.005	0.4%
AAG	0.003	0.25%

[a,b]See footnotes a and b, respectively in Table I

REFERENCES

Bolivar, F., Rodriguez, R.L., Greene, P.J., Betlach, M.C., Heyneker, H.L., Boyer, H.W., Crosa, J.H., Falkow, S. (1977) Gene 2, 95-113.

Bradford, M.M. (1973) Anal. Biochem. 72, 248-252.

Clark, B.F.C. and Marcker, K.A. (1966) J. Mol. Biol. 17, 394-406.

Dubnau, D. (1984) Crit. Rev. Biochem. 16, 103-132.

Duvall, E.J., Williams, D.M., Lovett, P.S., Rudolf, C., Vasantha, N., Guyer, M. (1983) Gene 24, 170-177.

Duvall, E.J., Ambulos Jr., N.P., Lovett, P.S. (1987) J. Bacteriol. 169, 4235-4241.

Gold, L. (1988) Ann. Rev. Biochem. 57, 199-233.

Gold, L. and Stormo, G. (1987) in Neidhardt, F.C., Ingraham, J.L., Magasanik, B., Low, K.B., Schaechter, M., Umbarger, H.E. (Eds), Escherichia coli and Salmonella typhimurium: Cellular and Molecular Biology. American Society for Microbiology, Washington, D.C. 1302-1307.

Hui, A. and De Boer, H. (1987) Proc. Natl. Acad. Sci. USA 84, 4762-4766.

Jacob, W.F., Santer, M. Dahlberg, A.E. (1987) Proc. Natl. Acad. Sci. USA 84, 4757-4761.

Kozak, M. (1983) Microbiol. Rev. 47, 1-45.

Laredo, J., Wolff, V., Lovett, P.S. (1988) Gene 73, 209-214.

Lovett, P.S. (1990) J. Bacteriol. 172, 1-6.

McLaughlin, J.R., Murray, C.L., Rabinowitz, J.C. (1981) J. Biol Chem. 256, 11283-11291.

Ray, C., Hay, R.E., Carter, H.L., Moran Jr., C.P. (1985) J. Bacteriol. 163, 610-614.

Reddy, P., Peterofsky, A., McKenney, K. (1985) Proc. Natl. Acad. Sci. USA 82, 5656-5660.

Roberts, M.W. and Rabinowitx, J.C. (1989) J. Biol. Chem. 264, 2228-2235.

Sharrock, W.J., Gold, B.M., Rabinowitz, J.C. (1979) J. Mol. Biol. 135, 627-638.

Shaw, W.V. (1975) Methods Enzymol. 43, 737-755.

Shine, J. and Dalgarno, L. (1974) Proc. Natl. Acad. Sci. USA 71, 1342-1346.

Tinoco Jr., I Borer, P.M., Dengler, B., Levine, M.D., Uhlenbeck, O.C., Crothers, D.M., Gralla, J. (1973) Nature New Biol. 246, 40-41.

Zukowski, M.M., Gaffney, D.F., Speck, D., Kauffmann, M., Findeli, A., Wisecup, A., Lecocq, J.-P. (1985) Proc. Natl. Acad. Sci. USA 80, 1101-1105.

MOLECULAR CLONING OF S-LAYER PROTEIN GENE FROM *BACILLUS THURINGIENSIS*

Chanpen Wiwat, Watanalai Panbangred,
Somsak Pantuwatana, Skorn Mongkolsuk and
Amaret Bhumiratana

Department of Biotechnology, Faculty of Science,
Mahidol University, Rama VI Road, Bangkok 10400
Thailand

ABSTRACT

Extraction of S-layer protein by treatment with a 6M urea indicated that there appeared to be extra-high molecular weight protein in the extracts obtained from *Bacillus thuringiensis* subsp. *israelensis (B.t.i.)* strain 4Q2. The antibody toward this large molecular weight protein was prepared and used for locating of S-layer protein on *B.t.i.* by using indirect immunofluorescent technique. Immunodiffusion reaction and Western blot analysis confirmed the specificity of the anti-S-layer protein.

It was found that the antibody against S-layer protein inhibited the plasmid transfer via the conjugation-like process. The frequency of transfer of plasmid pBC16 was found to be 9.7×10^{-6} and less than 10^{-8} in the presence and absence of anti-S-layer antibody, respectively. Using antibody detection technique, S-layer protein gene from *B.t.i.* strain 4Q2 was cloned in phagemid pKSII and plasmid pUC12 of *E.coli* DH5α. Three positive clones containing the gene encoding for the S-layer protein were obtained. Two clones had the insert of 5.2 kb, and the other one had an insert of 6.6 kb. The expression of the S-layer protein gene was detected by using Western blot analysis.

INTRODUCTION

Bacillus thuringiensis is one of the most effective bacterial insecticides and it is being used widely to control population of insect pests and vectors (1,2,3) *B. thuringiensis* toxins are highly selective in modes of action, such that the toxic activity of a particular isolate may be limited to very specific target insects (1,2).

Presently there are numerous attempts in trying to genetically improve strains of *B.thuringiensis*. The studies are either undertaken by cloning, characterizing and manipulating of toxin genes from *B.thuringiensis* via mutagenesis or gene transfer techniques (4,5,6,10,11,12).

Biotechnology and Environmental Science: Molecular Approaches
Edited by S. Mongkolsuk et al., Plenum Press, New York, 1992

In order to widen the host range of *B. thuringiensis* strain by gene manipulation, one of the most effective mean is conjugation-like process (7,8,9). Our previous study on the nature of intersubspecific gene transfer among *B. thuringiensis* have shown that the various *B.thuringiensis* subspecies could acquire the plasmids from *B.thuringiensis* subsp. *israelensis* which assigned as donor strain (13). The gene transfer via conjugation-like process required cell to cell interaction (14) and may thus involve on bacterial cell surface.

S-layer proteins have been observed on numerous species of gram-positive and gram-negative bacteria (15,16,17,28) their functions are not well known, and not well understood, the variety of functions including as a promoter for cell adhesion, as a supporting framework involved in maintaining the cell shaped, and as a barrier against external or internal factors have been proposed (15,17,18,19,20). Due to their location, S-layers may evolve as a chanel for gene transfer between bacterial cells (19).

In this study, the possible role of S-layer protein of *B.thuringiensis* subsp. *israelensis* on conjugalike-process was investigated. Furthermore, S-layer protein gene of *B.thuringiensis* was cloned in *E.coli* and the gene was characterized with the view to understand the detailed involvement of S-layer protein on conjugation-like gene transfer process.

MATERIALS AND METHODS

Bacterial Strains

Bacillus thuringiensis subsp. *israelensis* strains 4Q2, 4Q2-72 and -c.4Q2-72 were kindly provided by Dr. D.H. Dean, *Bacillus* Stock genetics center, the Ohio State University, Ohio, USA. Other strains are from our own stock culture collection.

The gene transfer using conjugation like process and the effect of anti 4Q2-S-layer protein

Conjugation-like process among *B. thuringiensis* subsp *israelensis* was performed by broth mating technique (13) described earlier by Wiwat et. al.

Penicillin G susceptibility testing

Penicillin G susceptibility of *B. thuringiensis* subsp. *israelensis* was performed by Kirby-Bauer disk diffusion method (21) and then determined the minimal inhibition concentration (MIC) by broth dilutions technique (21). Mueller Hinton medium (Difco) was used as testing medium.

Determination of Cell aggregation

Different strains of *B. thuringiensis* subsp. *israelensis* was separately grown in LB-broth in the same conditions as described for the conjugation-like procedure. The mixture of the two strains of *B. thuringiensis* subsp. *israelensis* was placed at room temperature for few minutes, the cell clumping was observed by macroscopic examination.

Molecular weight determination

Molecular weights were determined by running the proteins and molecular weight markers on 6.5% separating gels as described by Laemmli (22).

Preparation of antisera

Antisera were raised against S-layer protein in New Zealand white rabbits. The protein content was measured by Bradford method (23). Rabbits were immunized subcutaneously with 100 µg of S-layer protein which was suspended in 0.5 ml of sterile, distilled water and mixed with Freund complete adjuvant on days 1, and mixed with Freund incomplete adjuvant on days 14, and days 21. On days 28, blood was collected, the antibody titer was determined by Ouchterlony immunodiffusion. The serum was frozen at -20ºC.

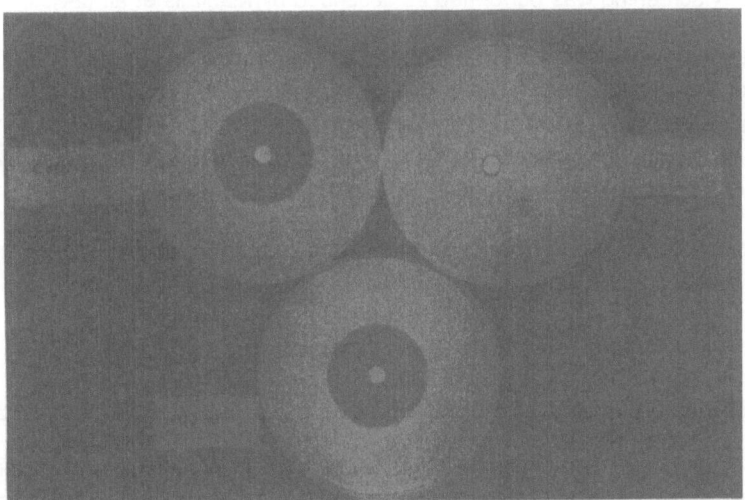

Fig. 1 Penicillin G susceptibility of various strains of *B.thuringiensis* subsp. *israelensis* determined by Disk Diffusion method. The cultures 4Q2, 4Q272, and c4Q272 were spreaded on Mueller-Hinton agar plates, penicillin G disks were placed on the spreaded plates and incubated overnight. This figure showed inhibition zones surrounded strains 4Q272 and c4Q272 (penicillin G susceptible), and no inhibition zone surrounded strain 4Q2 (penicillin G resistant).

Western Immunoblotting

SDS-PAGE was performed, the electrophoretic transfer to PVDF membrane (Immobilon- P, Millipore, USA) by using Bio-Rad Mini blot apparatus at 100 V for 2 hr. Portion of the PVDF membrane were stained briefly with 1% amido black, destained with 10% acetic acid, and 30% methanol, rinsed with distilled water, and allowed to air dry. Another portions of the PVDF membrane were blocked with 3% Bovine serum albumin (Sigma), rinsed with PBS pH 7.1 and treated with anti S-layer protein antiserum for 2 hr at room temperature, washing 3 times in PBST at 10 min each time and then incubated with goat anti-rabbit alkaline phosphatase (Sigma) diluted 1:1000 in PBST containing 3% Bovine serum albumin for 1 hr at room temperature. Washing the membrane as described above, and added substrates, the mixture of o-dianisidine tetrazotized (sigma) and beta- naphthyl phosphate (sigma) in substrate buffer (0.1 M carbonate buffer containing 1mM $MgCl_2$). The reactions were stopped by rinsing with water.

Cloning of S-layer protein gene

B. thuringiensis subsp. *israelensis* strain 4Q2[Nal] DNA was isolated, partially digested with the restriction enzymes, *Bam*HI, *Cla*I *ECo*RI, *Hin*dIII, *Pst*I and *Sau*3A, and then ligated into Bluescript phagemid vector pKSII or pUC12 cut with the same restriction endonuclease, *E.coli* DH5α was transformed and recombinant clones were identified on LB-agar plates containing ampicillin (100 µg/ml), IPTG and X-gal. The 5.2 kb fragment of *Pst*I-*Pst*I was subcloned in pKSII.

Immunological screening

Colony screening was performed as described by Maniatis et al. (24).

Protein immunoblotting was performed by using 3 ml of overnight culture of *E.coli*, centrifuge for 1 min in eppendorf tube at 7000 rpm and suspend in 85 µl distilled water. Then 100 µl of lysis buffer was added to each sample. Samples were heated in boiling water for 5-10 min, 5 µl each of sample was mixed with 5 µl of SDS-PAGE sample buffer (22), boiled for another 3-5 min and then was loaded on 1 mm thick SDS-PAGE.

Protein was blotted on nitrocellulose membrane and performed immunological reaction as previously described.

RESULTS

I. Cell aggregations between various strains of *B.thuringiensis* subsp. *israelensis*

During experiments on conjugation-like gene transfer process, it was observed that when certain pairs of *B.thuringiensis* subsp. *israelensis* were mixed together, cell aggregations were observed. The mutant strains derived from the same wild type strain (4Q2) can be devided into two groups, namely, group I (composed of strains 4Q2, c.4Q2-72-16 which harboured plasmid pBC16 of *B.cereus* GP7 by conjugation-like process) and group II (composed of strains 4Q2-72 and c.4Q2-72).

When *B.t.i* strains from group I were mixed with one of the strains from group II clumping occur, however, there is no clumping when the strains from the same grouping were mixed together. There was no self clumping in all the strains used in these experiments. The data as summarized in Table1 indicated that there might be a common factor(s) presence or missing from strain 4Q2-72 and c.4Q2-72 during the curing experiment which lead to the clumping phenomenon. However, this common factor is regained by the transfering of pBC16 plasmids in strain c.4Q2-72.

There was no clumping between the culture filtrate of strain 4Q2 or c.4Q2-72-16 and cells or culture filtrate of strain 4Q2-72 or c.4Q2-72, and vice versa (data was not shown). Therefore, there appeared to be requirement for the presence of two different types of cells for initiation of clumping phenomenon. Moreover, the growth of cells in different culture media; LB broth, TSB, NB and BHI, did not appear to effect the ability of cells to clump.

II. Penicillin susceptibility

It has been well documented that most wild type strains of *B.thuringiensis* subsp. *israelensis* possessed resistances to penicillin. The penicillin susceptibiliy test using disk

Table 1. Clumping between various strains of *B. thuringiensis* subsp. *israelensis*.

B.t.i. strain	4Q2	4Q2-72	c.4Q2-72	c.4Q2-72-16
4Q2	-	++++	++++	-
4Q2-72	++++	-	-	++++
c.4Q2-72	++++	-	-	++++
c.4Q2-72-16	-	++++	++++	-

diffusion method (Fig. 1) and broth dilution method demonstrated that the wild type strain 4Q2 was found to be highly resistant to penicillin G, i.e. minimal inhibitory concentration of 12,500 μg. However strains derived from 4Q2, i.e. strains 4Q2-72, c.4Q2-72 and c.4Q2-72-16 were found to be quite susceptible to penicillin at very low concentration with MIC of 0.06, 0.03 and 0.02 μg respectively.

III. The effect of anti-4Q2 S-layer protein antibody on conjugation-like process

The antibody raised against S-layer protein of *B. thuringiensis* subsp. *israelensis* strain 4Q2 (as shown in Table 2) could be shown to inhibit the conjugation-like gene transfer between *B.thuringiensis* subsp. *israelensis* strain 4Q2-16, harboring plasmid pBC16 and strain c.4Q2-72 (rif^R). This mating experiment was demonstrated by detection the transconjugants which acquired the plasmid pBC16. The control experiments, the conventional broth mating technique and mating in phosphate buffer medium clearly showed that the plasmid pBC16 could be transferred at frequency of transfer of 1.6×10^{-6} and 9.7×10^{-6}, respectively.

Also, the mating mixture which pretreated the donor strain with preimmunize rabbit serum or heated anti-4Q2 S-layer protien antiserum only slightly inhibited the transfer of pBC16 plasmids. The frequency of transfer was found to reduced to 1.9×10^{-6} and 0.1×10^{-6} with the treatment of donor strain with preimmunized rabit serum as shown in Table 2.

Table 2 The effect of anti-4Q2 S-layer protein antibody on conjugation-like process between *B. thuringiensis* subsp. *israelensis* strains 4Q2-16 and c.4Q2-72.

condition of conjugation		Freq. of
medium	donor	transfer[a]
LB-broth	no treatment	1.6×10^{-6}
phosphate buffer	no treatment	9.7×10^{-6}
phosphate buffer	treated with pre-immunized rabbit serum	1.9×10^{-6}
phosphate buffer	treated with anti-4Q2 S-layer protein antibody	less than 10^{-8}
phosphate buffer	treated with heated anti-4Q2 S-layer protein antibody	0.1×10^{-6}

a) average of three independent experiments.
 calculated by dividing number of transconjugants by number of donor.

IV Molecular Cloning of the S-layer protein gene from *B.t.i.*

A genomic DNA library constructed in phagemid vector pKSII and plasmid vector pUC12 from a partial digestion of the whole DNA preparation of *B.t.i.* 4Q2[Nal] by various enzymes as described in materials and methods, was screened with the anti 4Q2-antibody and alkaline phosphatase conjugated with anti rabbit-IgG.

An antibody to the S-layer protein of *B.t.i.* strain 4Q2 was used as antibody probe for cloning of the *B.t.i.* 4Q2 high molecular weigth protein gene. In order to avoid non-specific interaction, the anti-4Q2 S-layer protein antibody was adsorbed with *E.coli* harboring only pKS-II or pUC12 vector. Approximately 30,000 colonies of *E.coli* DH5α tranformants were screened, three positive clones designated pAB1, pAB2, pAB3 were obtained and found to contain 6.6-kb insert in pAB1 clone, 5.2-kb inserts in pAB2 and pAB3 as shown in Fig 4. The presene of S-layer protein cloned gene products were confirmed by Western blot analysis as shown in Fig. 3.

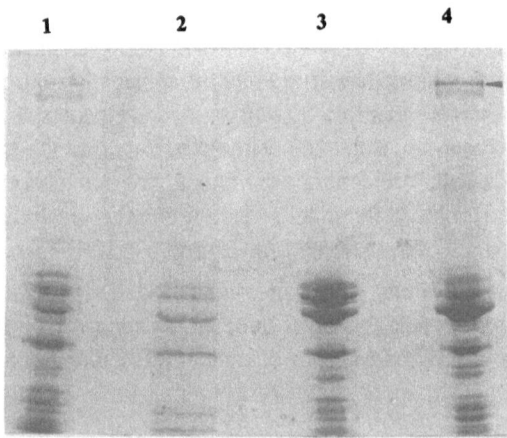

Fig.2   SDS-PAGE patterns of protein extract of *B. thuringiensis* subsp. *israelensis.* The crude protein extracts of strains c4Q272-16(lane 1), c4Q272(lane 2), 4Q272(lane 3) and 4Q2 (lane 4) were prepared and subjected to SDS - PAGE.

DISCUSSION

Our study has shown that there appeared to be some correlation between the presence and absence of the high molecular weight protein, i.e. S-layer protein with penicillin G susceptibility. The wild type strain, 4Q2, possesses the high molecular weight, S-layer protein and it is very highly resistant to penicillin G, at MIC approximately 12.5 mg both of the cured strain 4Q2-72 which harbors only one plasmid and the plasmidless strain c.4Q2-72 which does not possess the S-layer protein are very sensitive to penicillin G at MIC 0.06 and 0.03 µg respectively. This phenomena seemed to suggest that, some factors which might be loss or gained during the curing process, and these factors might be involved in the changes in susceptibility to penicillin G among strains of *B. thuringiensis* subsp. *israelensis*. This evidence was supported by some previous studies which have been reported that several mesophilic Bacillaceae possessing S-layer are resistant to lysozyme (16,17,19) and also S-layer on intact cells are frequently rather resistant to a wide spectrum of protease (15).

Due to the cell clumping between various strains of *B. thuringiensis* sub. *israelensis* which we categorized into group I and group II. There appeared to be some common factors within the same group, those are, the presence of S-layer protein of group I and the absence of S-layer protein of group II. Interestingly, the clumping occurred when cell of group I was mixed with cell of group II, it might be interpreted that the clumping was only resulted from the interaction between S-layer protein on cell surface and uncoated cell. This phenomenon might be similar to the presence of S-layer protein on *Aeromonas salmonicida* decreases surface hydrophobicity and thus enhances the ability for autoagglutination and for association with phagocytic monocytes (29,30).

Using the antibody raised against S-layer protein of *B.thuringiensis* subsp. *israelensis* strain 4Q2 could be shown to inhibit the conjugation-like process between *B. thuringiensis* subsp. *israelensis* strain 4Q2-16 and c.4Q2-72 at the detectable level of transconjugant by broth mating technique. This experiment demonstrated that S-layer protein might play a role on conjugation-like gene transfer process, therefore when there was anti-S-layer protein

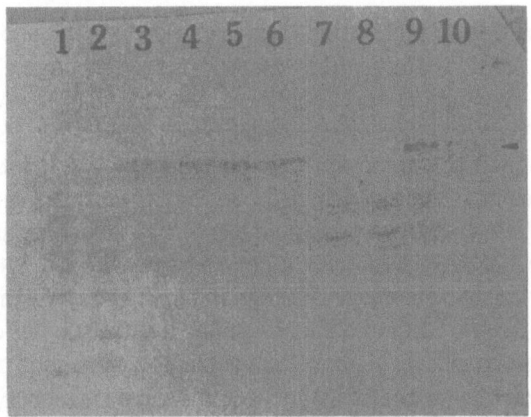

Fig. 3 Western blotting of protein extracts from *E.coli* DH5α transformants. Protein bands from *E.coli* harboring pUC12 vector (lanes 1 and 2), *E.coli* harboring pAB2 (lanes 3 and 4), *E.coli* harboring pAB3 (lanes 5 and 6), *E.coli* harboring pBluescript KS vector (lanes 7 and 8), and *E.coli* harboring pAB1 (lanes 9 and 10) were transferred onto blotting membrane after electrophoresis on SDS-PAGE. And then detection of positive band was carried out by using anti - 4Q2 S-layer protein antibody as primary antibody.

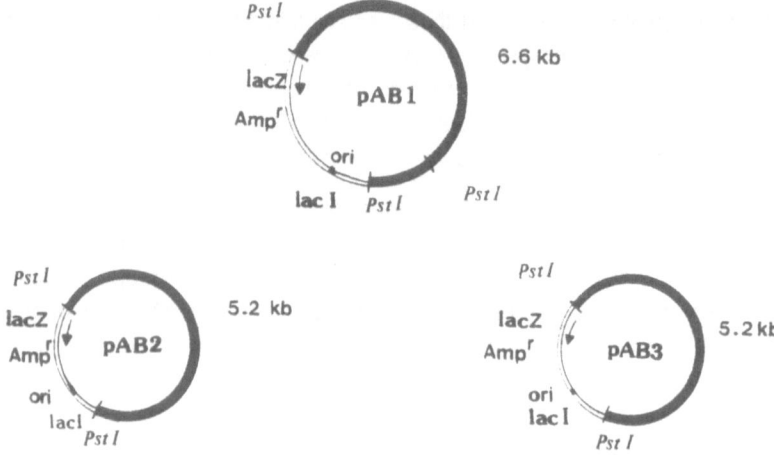

Fig.4 The positive clones of S-layer protein gene. pAB 1. contained 6.6 kb *Pst* I fragment, pAB2 and pAB3 contained 5.2 kb *Pst* I fragment.

antibody block on cell surface, the gene transfer via conjugation-like process was also inhibited. There is good evidence from 3-dimensional reconstructions as well as from higher resolution projection data (31), that S-layer proteins might cause 'connecon' chanel with an average diameter of about 2.5 nm (19) which might be sufficient for DNA transfer via conjugation-like process. Therefore, the antigen-antibody interaction between S-layer and anti-S-layer might block or shield this chanel and then inhibit DNA transfer.

In this study, the S-layer protein gene has been cloned from genomic DNA library of *B.thuringiensis* subsp. *israelensis* strain 4Q2[Nal] and constructed in phagemid vector pKSII and plasmid vector pUC12, transformed into *E. coli* DH5α, and screening positive clones with anti-4Q2-antibody probe. Three positive cloned containing 6.6 kb and 5.2 kb *Pst*I insert fragments were obtained. The cloned gene products were confirmed by Western blot analysis and it showed the protein at same molecular weight. However, the molecular weight of the cloned gene products in *E.coli* are lower than gene products of the strain 4Q2, and also lower than the expected gene product calculated from the 5.2 kb fragment gene. The different size of *Pst*I fragment, 6.6 kb and 5.2 kb gene expressed the same molecular weight of gene products, therefore, it might be assume that the 5.2 kb fragment is necessary for coding the S-layer protein gene, and it may be incomplete gene due to the size of its gene product. However, the molecular weight was determined by SDS-PAGE, therefore the mobility of S-layer protein might interfere with carbohydrate moiety or lipid moiety attach to the protein molecule and also the proteolytic cleavage in *E. coli* may reduce the molecular size. More experiments on these aspects will be needed for further clarification.

There are few reports on cloning of S-layer protein genes (32,33,34,35), all of these genes could be expressed in *E. coli* host. And also the DNA fragment in transformed *E. coli* coded for S-layer protein of *B. licheniformis* NM 105 had a lower molecular mass than the authentic S-layer protein (33) which was suggested to be due to protease degradation. In addition, all reports demonstrated that the S-layer protein gene were located on chromosomal

DNA (32,33,34,35), however the S-layer protein gene from *B. thuringiensis* subsp. *israelen sis* strain 4Q2-16 which could transfer to c.4Q2-72 (Table 1 and Fig. 2) provide preliminary suggestion that this gene might be located on plasmid rather than chromosomal DNA.

The detailed study of S-layer protein gene with regard to further characterization and determination of its DNA sequence are in progress.

REFERENCES

1. Andrews, R.E. and L.A. Bulla. 1982. Toxins of sporeforming bacteria. In : Spore III (Levinson, H.S., A.L.Sonenshein and D.J. Tipper, eds.) p. 57, American Society for Microbiology, Washington DC.
2. Aronson, A.I., W.Beckman and P. Dunn. 1986. *Bacillus thuringiensis* and related insect pathogens. Microbiol. Rev. 50: 1-24.
3. Martin, P.A.W. and D.H. Dean. 1981. Genetics and genetic mainpulation in *Bacillus thuringiensis*. In: Microbial Control of Pests and Plant Diseases 1970-1980 (Burges, H.D., ed.). p. 299, Academic Press, London.
4. Fischer, H.M., P.Luthy and S.Schweitzer. 1984. Introduction of plasmid pC194 into *Bacillus thuringiensis* by protoplast transformation and plasmid transfer. Arch. Micro-biol. 139: 213-217.
5. Martin, P.A.W., J.R. Lahr, and D.H. Dean. 1981. Transformation of *Bacillus thuringiensis* protoplasts by plasmid deoxyribonucleic acid J. Bacteriol 145: 980-983.
6. Miteva, V.I., N.I.Shivarova, and R.T. Grigorova. 1981. Transformation of *Bacillus thu-ringiensis* protoplasts by plasmid DNA. FEMS Microbiol. Lett 12: 253-256.
7. Gonzalez, J.M., B.J. Brown and B.C. Carlton. 1982. Transfer of *Bacillus thuringiensis* plasmids coding for Delta-endotoxin among strains of *B. thuringiensis* and *B. cereus*. Proc. Natl. Acad. Sci. U.S.A. 79: 6951-6955.
8. Klier, A., C. Bourgouin and G.Rappoport. 1983. Mating between *Bacillus subtilis* and *Bacillus thuringiensis* and transfer of cloned crystal genes. Mol. Gen. Genet. 191: 257-262.
9. Loprasert, S., Pantuwatana and A. Bhumiratana. 1986. Transfer of plasmids pBC16 and pC194 into *Bacillus thuringiensis* subsp *israelensis* J. Invert. Pathol. 48: 325-334.
10. Mahillon, J., W. Chungjatupornchai, J. Decock, S. Dierickx, F. Michiels, M. Peferoen and H.Joos.1989. Transformation of *Bacillus thuringiensis* by electroporation. FEM Microbiol Lett 60: 205-210.
11. Bone, E.J., and D.J. Ellar. 1989. Transformation of *Bacillus thuringiensis* by electropo-ration. FEMS Microbiol Lett 58: 171-178.
12. Lereclus, D., O. Arantes, J. Chaufauxe, and M.-M. Lecadet. 1989 Transformations and expression of a cloned Delta-endotoxin gene in *Bacillus thuringiensis*. FEMS Microbiol Lett 60: 211-218.
13. Wiwat, C., W. Panbangred, and A. Bhumiratana. 1990. Transfer of plasmids and chro-mosomal genes amongst subspecies of *Bacillus thuringiensis* J. Ind. Microbiol. 6: 19-27.
14. Chapman, J.S. and B.C. Carlton. 1985. Conjugal Plasmid transfer in *Bacillus thuringiensis*. In Plasmid in Bacteria (Helinski, D.R., S.N. Cohen, D.B. Clewell, D.A.Jackson and A. Hollaender, eds.) pp. 453-467. Plenum Press, New York.
15. Sleytr, U.B. and P.Messner. 1988. Crystalline surface layers in procaryotes. J. Bacteriol. 170: 2891-2897.

16. Koval, S.F., and R.G.E. Murray 1984. The isolation of surface array proteins from bacteria. Can. J. Biochem. Cell Biol. 62: 1181-1189.

17. Sleytr, U.B., and P. Messner. 1983. Crystalline surface layers on bacteria. Annu. Rev. Microbiol 37: 311-339.

18. Sleytr, U.B. and M. Sara. 1986. Ultrafiltration membranes with uniform pores from crystalline bacterial cell envelope layers. Appl. Microbiol. Biotechnol. 25: 83-90.

19. Baumeister, W. and R. Hegerl. 1986. Can S-layers make bacterial connexons. FEMS Microbiol Lett 36: 119-125.

20. Beveridge, T.J. 1988. The bacterial surface: general considerations towards design and function. Can. J. Microbiol. 34: 363-372.

21. USPXX. 1980. The Pharmacopocia of the United States of America. The United States Parmocopoeial Convention, Ltd. Rockville, Md., USP.

22. Laemmli, U.K. 1970. Cleavage of structural proteins during the assembly of the head of bacteriophage T4. Nature 227: 680-685.

23. Perbal, B. 1988. A practical guide to molecular cloning. 2nd Ed. p.49. John wiley & sons., New York.

24. Sambrook, J., E.F.Fritsch, and T. Maniatis, 1989. Molecular Cloning : A Laboratory manual 2nd Ed. Cold Spring Harbor Laboratory Press. New York.

25. Ellar, D., D.G. Lundgren. 1967. Ordered substructure in the cell wall of *Bacillus cereus*. J. Bacteriol. 94: 1778-1780.

26. Wallinder, I.B., and H.Y. Neujahr. 1971. Cell wall and peptidoglycan from *Lactobacillus fermenti* J. Bacteriol 105: 918-926.

27. Beveridge, T.J. 1981. Ultrastructure, Chemistry and function of the bacterial wall. Int. Rev. Cytol 72: 229-317.

28. Luckevich, M.D., and T.J. Beveridge. 1989. Characterization of a dynamic S-layer on *Bacillus thuringiensis* J.Bacteriol 171:6656-6667.

29. Evenberg, D., and B. Lugtenferg. 1982. Cell surgace of the fish pathogenic bacterium *Aeromonas salmonicida* II. Purification and characterization of a major cell envelope protein related to autoagglutination, adhesion and virulence. Biochem. Biophys. Acta 684. 249-254.

30. Trust, T.J., W.W. Kay, and E.E. Ishiguro. 1983. Cell surface hydrophobicity and macro-phage association of *Aeromonas salmonicida*. Curr. Microbiol. 9: 315-318.

31. Baumeister, W., M. Barth, R. Hegerl and R.Guckenberger. 1986. Three-dimersional structure of the regular surface layer (HPI layer) of *Deinococcus radiodurans*. J. Mol. Biol. 187: 241-253.

32. Yamagata, H., T. Adachi, A. Tsuboi, M.Takao, T.Sasaki, N.Tsukagoshi and S. Udaka. 1987. Cloning and characterization of the 5' Region of the cell wall protein gene operon in *Bacillus brevis* 47. J. Bacteriol. 169: 1239-1245.

33. Tang, M., K. Owens, R. Pietri, X. Zhu, R. Meveigh, and B.K. Ghosh. 1989. Cloning of the crystalline cell wall protein gene of *Bacillus licheniformis* NM105 J. Bacteriol 171: 6637-6648.

34. Peters, J., M. Peters, F. Lottspeich, W. Schaffer and W. Baumeister. 1987. Nucleotide sequence analysis of the gene encoding the *Deinococcus radiodurans* surface protein, derived amino acid sequence and complementary protein chemical studies. J. Bacteriol. 169: 5216-5223.

35. Belland, R.J., and T.J. Trust. Cloning of the gene for the surface array protein of *Aeromonas salmonicida* and evidence linking loss of expression with genetic deletion. J. Bacteriol 169: 4086-4091.

36. Polak, J., R.P.Novick. 1982. Closed related plasmids from *Staphylococcus aureus* and soil bacilli. Plasmid 7: 152-162.

37. Bernhard, K., H. Schrempe, and W. Goebel. 1978. Bacteriocin and antibiotic resistance plasmids in *Bacillus cereus* and *Bacillus subtilis*. J. Bacteriol 133: 897-903.

38. Hanahan, D. 1985. Techniques for transformation of *E.coli* DNA cloning. In : DNA cloning vol 1. (Glover, D.M. ed.) IRL press, Oxford.

PRODUCTION OF 6-AMINOPENICILLANIC ACID BY AN ENZYMATIC PROCESS

Watanalai Panbangred, Saiyavit Varavinit,
Taweerat Vichitsoonthonkul, Suang Udomvaraphunt,
Suthum Intrararuangsorn and Vithaya Meevootisom

Department of Biotechnology, Faculty of Science,
Mahidol University, Rama VI Rd., Bangkok 10400, Thailand

6-APA can be produced by converting penicillin-G into 6-APA and phenylacetic acid with penicillin acylase (PA) enzyme. The genes encoding PA were cloned from Escherichi a coli and Bacillus megaterium. E. coli transformants containing the E. coli PA gene produce around 10 times higher enzyme activity than the parent cells. B. subtilis transformants containing the B. megaterium PA gene produce around twice the enzyme activity of the parent cells. Restriction maps and sizes of the two PA genes are different. Attempts to increase gene expression by using strong promoters were partially successful.

The E. coli transformant (pMLV6) has been immobilized on cotton threads with calcium alginate, resulting in a relatively small loss in enzymatic activity. The immobilized cells were quite stable when kept at 4°C and/or occasionally used (with less than 10% loss in activity) over a period of 5 months. When used for converting penicillin-G (15 liters of 6-8% commercial crude penicillin-G), 5 Kg of immobilized cells were needed to give a conversion of around 98% in 3-4 hours at 35°C. Crystallization of the 6-APA produced was achieved by using 33-50% methanol, lowering the pH to 4.2 and cooling to -5°C. Percent recovery of the crystallized 6-APA obtained was around 70-80%.

INTRODUCTION

6-Aminopenicillanic acid (6-APA) is a key intermediate for synthesis of most penicillin derivatives (1). Production of 6-APA is achieved by means of converting penicillin-G to 6-APA by the enzyme penicillin acylase (PA) (2). Escherichia coli and Bacillus megaterium are commonly used as sources of the enzyme in many studies (3-8). The enzyme is used in the forms of immobilized cells or immobilized enzyme in the production process. This is to reduce the catalyst cost due to the nature of the immobilized material which is reusable and stable (9) . In general, immobilized cells are easily prepared and less expensive than immobilized enzyme. However, they may have some unwanted side-reactions and are lower in activity per unit volume, consequently, they require larger bioreactors or longer operation time for

Biotechnology and Environmental Science: Molecular Approaches
Edited by S. Mongkolsuk et al., Plenum Press, New York, 1992

191

similar production capacity. With recent recombinant-DNA technology, enzyme yield has increased considerably (10-12). This makes the costs of immobilized cells and immobilized enzyme lower and, hence, the cost of 6-APA production.

It is the purpose of this study to employ recombinant-DNA technology to increase penicillin acylase activity in E. coli and Bacillus and to employ immobilization techniques to immobilize cells or enzyme for use in the 6-APA production.

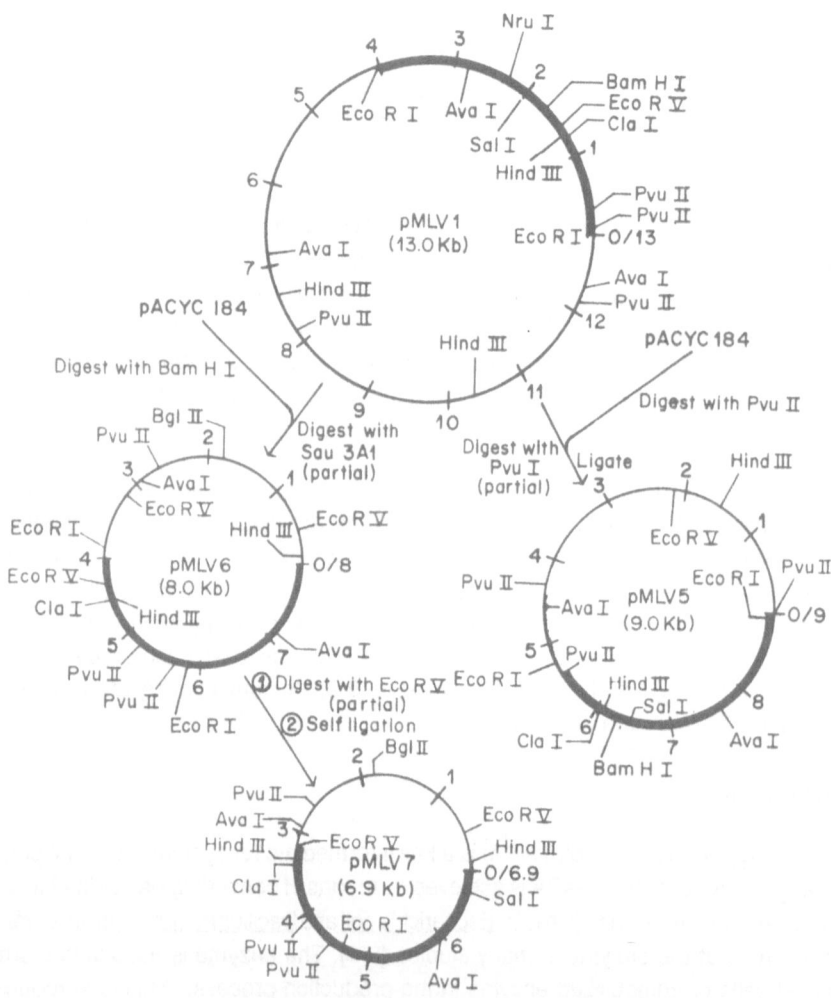

Figure.1 . Schematic presentation of plasmid constructions used to subclone the gene coding for penicillin G acylase from E. coli 194-3. Thick lines indicate plasmid vector (pACYC184) and thin lines indicate inserted fragment. Relevant restriction sites are indicated.

MATERIALS AND METHODS

Bacterial strains and plasmids. Escherichia coli 194 and 194-3 (13) and Bacillus megaterium UN1, Bacills subtilis MI 111 were from the collection of the Microbiology Department, Faculty of Science, Mahidol University. E. coli ATCC 9637 and Serratia marcescens ATCC 27117 were obtained from the American Type Culture Collection, Rockville, Md., USA. E. coli DH1 (14) and the plasmid pACYC184 (15) were from the collection of the Department of Microbiology, University of Liverpool. E. coli DH5α and the plasmid pTTQ18, pMLV6, pMLV7 pMLV101 and pBA401 were from the collection of the Department of Biotechnology, Faculty of Science, Mahidol University.

Media and reagents

Microbiological media were obtained from Difco. The standard media used in this work were L broth (LB) and L agar (LA). LB medium composed of 1% tryptone, 0.5% yeast extract, and 0.5% NaCl. LA had the same composition with LB plus 1.2% agar. Nalidixic acid, tetracycline, chloramphenicol, penicillin-G potassium salt, phenylacetic acid, 6-APA, and IPTG were purchased from Sigma Chemical Company (Miss., USA). Restriction enzymes were obtained from either BRL, Pharmacia, Takara, Toyobo or Sigma.

Assay for penicillin G acylase activity

Cells were grown in liquid medium containing bacto-peptone (0.5%), yeast extract (0.3%), and 0.15% PAA, pH 7.0. After 24 h shaking at 28°C, the cells were centrifuged and resuspended in 0.05 M Tris-HCl buffer, pH 7.5. Cell concentration was adjusted to an O.D. of 1.0 at 660 nm. Then 1.0 ml of this cell suspension was recentrifuged in a microcentrifuge tube and the sediment was used as an enzyme source in a reaction mixture containing 5 mg penicillin G, in 500 ul of Tris-HCl buffer (0.05 M, pH 7.5). The mixture was incubated at 42°C for up to 40 min and was then assayed for 6-APA by the p-dimethylaminobenzaldehyde method (16). When the culture broth of Bacillus culture was used instead of cells, 0.1 ml of the culture broth was used in the reaction mixture. Then, assay was similarly done as with E. coli cells as described previously.

Screening for penicillin G acylase-positive clones

Transformants grown on LA plates containing both the appropriate antibiotic (of final concentratiom conce concentrations, 10 ug/ml for chloramphenicol; or 50 ug/ml for nalidixic acid; or 15 ug/ml of kanamycin) and phenylacetic acid (0.15%), were then over-laid with 5 ml of soft agar containing Serratia marcescens (0.5 ml of an overnight culture per 100 ml) and K-penicillin G (5 mg/ml). The plates were incubated at 28°C for 48 h and positive clones were detected as previously described (10).

Recombinant DNA work

Plasmid DNA from E. coli and Bacillus was prepared as described by Birnboim and Doly (17) and by Udomvaraphunt (18), respectively. Restriction endonuclease digestion was done following the methods of Maniatis et al., (19). Other recombinant-DNA technology work including recovery of DNA fragment from agarose gel, plasmid transformation of competent E. coli or Bacillus and so on, were done as described by Udombunditkul (20) and Udomvaraphunt (18).

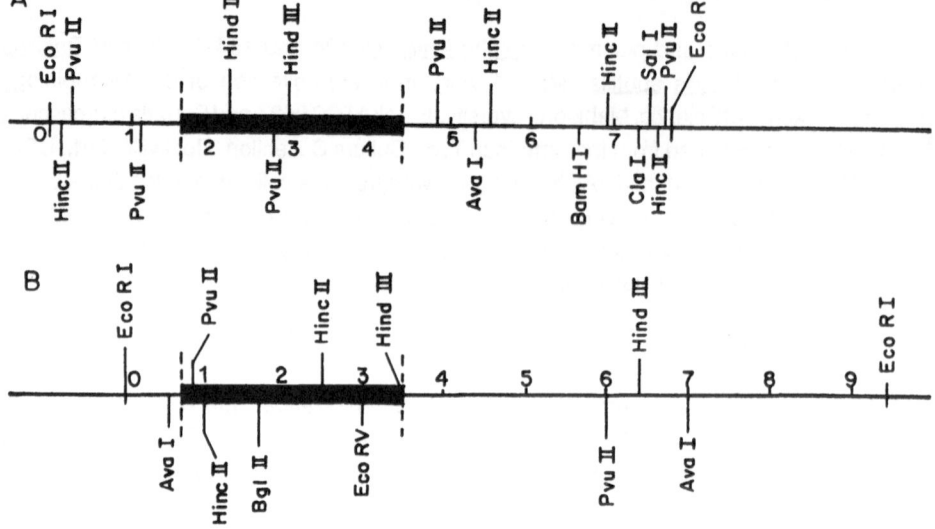

Figure. 2 Comparison of restriction maps of penicillin acylase genes from
B. megateium (A) and E. coli (B).

Immobilization of E. coli cells

One gram of cell paste of E. coli (pMLV6) was mixed with one ml of 4% sodium alginate containing 0.2% polyethyleneimine (PEI). The mixture was thoroughly mixed before being fixed on cotton cloth or cotton thread. The cell cloth or cell thread was then placed in 0.2 M CaCl$_2$ containing 0.5% glutaraldehyde for 30 minutes.

RESULTS AND DISCUSSION

Cloning of E. coli and Bacillus penicillin acylase genes

pACYC 184 was used as a vector in cloning both E. coli and Bacillus penicillin acylase (PA) genes. The vector and the bacterial chromosomal DNA were cut with EcoRI. The transformants containing chimeric plasmids pMLV1 and pMLV101 contained the PA genes of E. coli and B. megaterium, respectively. pMLV1 was used in subcloning experiments to obtain pMLV5, pMLV6 and pMLV7 as shown in Figure 1.

Comparison of penicillin acylase genes of E. coli and B. megaterium

Localization of the PA genes of both E. coli and B. megaterium was done by using Tn1000 insertion inactivation techniques (21). The genes were found to be around 2.5-2.7 Kb in size which were large enough to code for the E. coli PA enzyme of 70,000 daltons and for the Bacillus PA enzyme of 60,000 daltons. Comparison of retriction maps of both E. coli and Bacillus PA genes was found to be different from each other as shown in Figure 2.

194

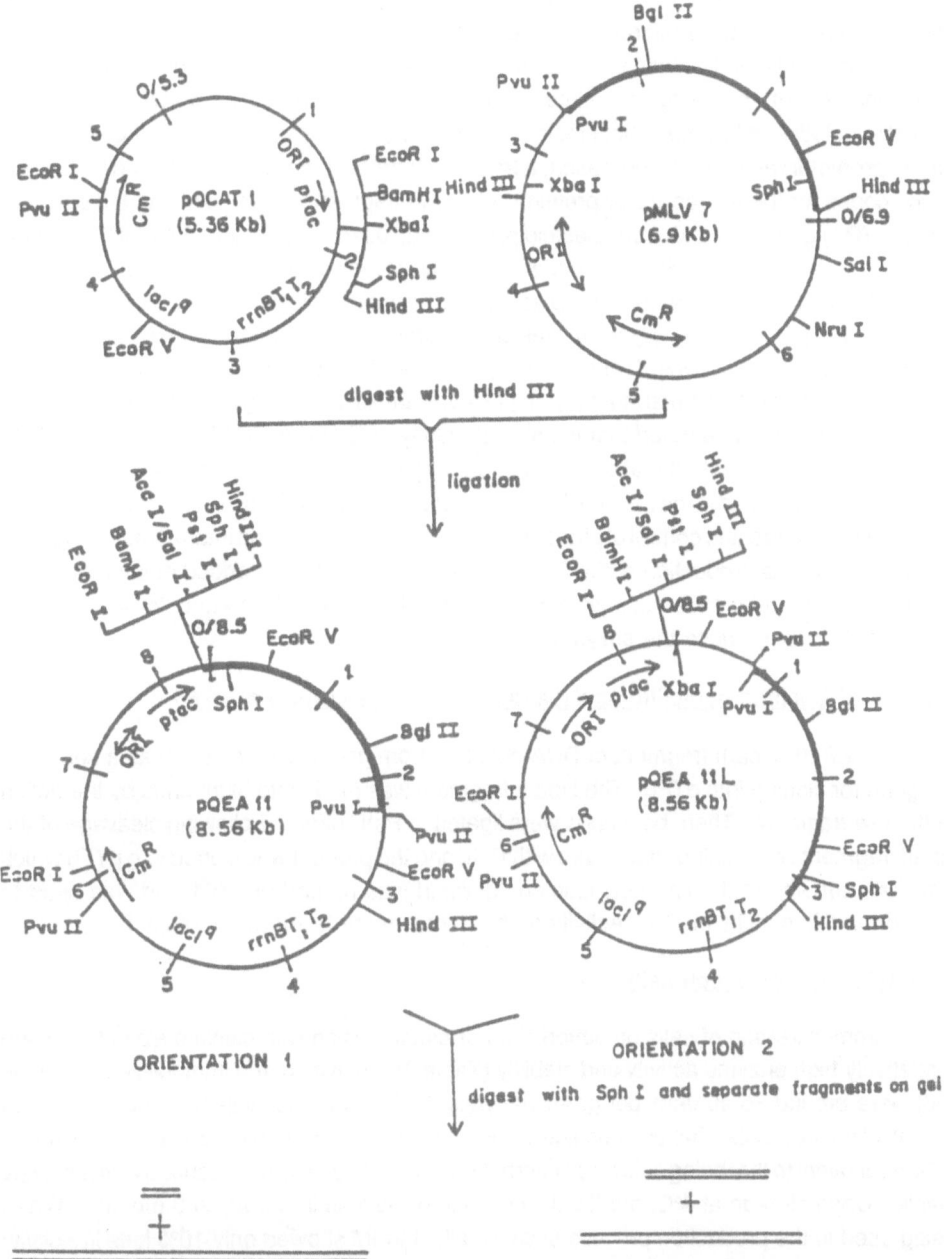

Figure 3. Schematic presentation of construction plasmid pQEA11 and determination of penicillin acylase gene insertion in pQCAT1. Thick lines indicate penicillin acylase gene and thin lines indicate plasmid vector (pQCAT1). The clone with correct orientation (orientation 1) of penicillin acylase provides large inhibition zone.

Cloning of E. coli penicillin acylase gene under the control of ptac

pTTQ18 was used as a starting vector. It was modified by inserting a promotorless chloramphenicol acetyltransferase (cat) gene into the beta-lactamase gene on pTTQ18 at the Scal site. This modification was required because the beta-lactamase enzyme interferes with transformant screening procedure. The new plasmid was called pQCAT1. Its main features includes chloramphenicol resistant gene as a selective marker; a ptac which is a strong promotor; an origin of replication; a transcriptional terminator, rrnBT$_1$T$_2$ and laclq gene which codes for the lac repressor protein. A Hind III fragment, which contains the penicillin acylase (PA) gene of 3.2 kb from plasmid pMLV7 was used to clone into the pQCAT1 at the HindIII site. This made the PA gene under the control of the ptac promotor as shown in Figure 3, orientation 1. This new chimeric plasmid was named pQEA11. When it was used to transform E. coli DH5α, the organism was able to produce large amount of PA enzyme after IPTG was used as an inducer (Table 1). E. coli (pQEA11) also showed larger clear inhibition zone compared to the parental strain, E. coli 194-3, or those containing pMLV6 or pMLV7 (data not shown). It should be noted that the size of colony of E. coli (pQEA11) was smaller than that of the others indicating that growth was partially inhibited. This was expected because IPTG was incorporated in the medium before inoculation of E. coli (pQEA11). Over expression of the PA gene under the control of ptac might have drained energy and intermediates required for growth towards production of the PA enzyme. When IPTG was added at later stages of growth in broth, culture biomass yield was better but IPTG has less effect for inducing the organism to over produce the enzyme.

Cloning of Bacillus penicillin acylase gene under the control of pTF6

A 2.8 Kb Hpa II fragment of DNA obtained from pMLV101 was used as a source of PA gene for cloning into pTF6. The Hpa II fragment was made into blunt ends by the action of Klenow fragment. Then, both ends were ligated to PstI linkers. Following cleavage of the HpaII fragments containing PstI linkers with PstI, and the product was cloned into pTF6 which was also cut with PstI. The new chimeric plasmid was named pBA401 and was used to transformed B. subtilis MI111. Activity of the transformant is shown in Table 1.

Immobilization of E. coli cells

Immobilization of cells on cotton cloth or cotton thread with calcium alginate yielded a relatively high enzyme activity and stability (Table 2). Activity of the immobilized materials (IM) were around 35-40 units per gram wet weight. This was equivalent to around 65% of activity of the free cells. The enzyme activity obtained with our immobilized enzyme was found to be equivalent to that being sold by an international company. The IM produced was relatively stable. Upon storage at 4°C, the IM showed little or no loss in activity in 3 months. When being used in the production process occasionally, the IM showed only 10% loss in activity in five months.

Production of 6-APA

When the immobilized cell thread was used in the production process, the IM could convert 6-8% of penicillin-G potassium salt to 6-APA within 3-4 hours with around 97%

conversion (Table 3). It is agreed that concentration of penicillin-G for use in the production process should be around 6-10% to be economically feasible. Lower concentration of penicillin-G would yield lower concentration of 6-APA which makes the downstream process costly. The time required in the conversion process has to be within 3-4 hours to minimize both substrate and product degradation since the nature of beta-lactam ring is relatively unstable. Moreover, the percent conversion of penicillin-G to 6-APA must be high, i.e., over 95% to allow effective crystallization of 6-APA to occur. Lower percent conversion yielded much lower percent of 6-APA recovery (data not shown).

Crystallization of 6-APA

At the end of the conversion process, the mixture was mixed with cold methanol at 33-50% final concentration. The mixture was mixed well and cooled down to -5°C to 0°C. Then 6N HCl was used to adjust the pH of the mixture to 4.2. After 3 hours of incubation with constant mixing 6-APA was mostly crystallized with around 78% yield and 95% purity. This was considered to be low in terms of percent recovery. In practice, in some industrial companies the yield was claimed to be around 86-87%. We believed that with improvement of the facilities used, especially, the cooling system and the pH controlling system we should be able to achieve the percent recovery close to 86-87% obtained by the others.

Table 1. Activities of penicillin acylase in producer strains and clones.

Strain	Penicillin acylase activity[a]	
	Uninduced[b]	Induced
E. coli ATCC9637	3	10
E. coli 194	0	6
E. coli 194-3	0	22
E. coli (pMLV1)	40	10
E. coli (pMLV6)	104	102
E. coli (pMLV7)	106	100
E. coli (pQEA11)	100	230
E. coli DH1	0	
E. coli DH5α	0	0
B. megaterium UN1	ND	0.67
B. subtilis (pBA401)	ND	1.26
B. subtilis MI111	ND	0

[a]Penicillin acylase activity is expressed in terms of units per gram cell wet weight (u/gcww) for E. coli and units per ml of culture broth for Bacillus.

[b]Uninduced and induced indicate that PAA was omitted from or added to the growth medium, respectively, for all organisms except E. coli (pQEA11) and E. coli DH5α where IPTG was used instead of PAA.

Table 2. Characteristics of immobilized cells.

Characteristic		
Enzyme activity	35-40	Units/gww
	(150-200)	Units/gdw)
Percent remaining activity	65%	
Stability		
Storage at 4°C	>3 months	
Operating (occasionally)*	>5 months	

*Operating 25 times with 3-4 hours each, and kept at 4°C during storage - only 10% loss in activity.

Table 3. Pilot scale production of 6-APA.

Characteristic	
1. Working volume of reactor	20L/0.5 Kg
2. Concentration of penicillin-G used	6-8%
3. Percent conversion of penicillin-G to 6-APA (in 3-4 hours)	97%
4. Time required for the conversion process	3-4 hours

ACKNOWLEDGEMENTS

This work was supported by the National Center for Genetic Engineering and Biotechnology and by Mahidol University.

REFERENCES

1. Sheehan, J.C., Henry-Logan, K.R.: J. Am. Chem. Soc. 84, 2983 (1962).
2. Savidge, T.A.: In Vandamme, E.J. (ed). Biotechnology of industrial antibiotics, Marcel Dekker Inc., N.Y. (1984).
3. Self, D.A., Kay, G., Lilly, M.D., Dunnill, P.: Biotechnol. Bioeng. 11, 337 (1969).
4. Rolinson, G.N., Batchelor, F.R., Butterworth, D., Cameron-Wood, J., Cole, M., Eustace, G.C., Hart, M.V., Richards, M., Chain, E.B.: Nature 187, 236 (1969).
5. Szentirmai, A.: Appl. Microbiol. 12, 185 (1964).
6. Park, J.M., Choi, C.Y., Seong, B.L., Han, M.H.: Biotechnol. Bioeng. 24, 1623 (1982).
7. Chiang, C., Bennette, R.E.: J. Bacteriol. 93, 302 (1967).
8. Theeragool, G.: Studies on penicillin acylase of Bacillus megaterium UN1 and some applictions for future industrial use, M.Sc. Thesis, Mahidol University, Bangkok, Thailand (1985).
9. Chibata, I.: Immobilized enzymes: Research and development, John Wiley & Sons, N.Y. (1978).
10. Mayer, H., Collins, J., Wagner, F.: In Weetal, H.H., Royer, G.P. (eds). Enzyme Engineering, 5, 61 (1980).

11. Meevootisom, V., Saunders, J.R.: Appl. Microbiol. Biotechnol. <u>25</u>, 372 (1987).
12. McCullough, J.E.: Biotechnol <u>1</u>, 879 (1983).
13. Meevootisom, V., Somsuk, P.: J. Sci. Soc. Thailand <u>9</u>, 143 (1983).
14. Hanahan, D.: J. Mol. Biol. <u>166</u>, 557 (1983).
15. Chang, A.C.Y., Cohen, S.N.: J. Bacteriol. <u>134</u>, 1141 (1978).
16. Balasingham, K., Warburton, D., Dunnill, P., Lilly, M.D.: Biochim. Biophys. Acta <u>276</u>, 250 (1972).
17. Birnboim, H.C., Doly, J.: Nucleic Acids Res. <u>7</u>, 1513 (1979).
18. Udomvaraphunt, S.: Genetic manipulation of penicillin acylase in <u>Bacillus</u> sp., M.Sc. Thesis, Mahidol University, Bangkok, Thailand (1989).
19. Maniatis, T., Fritsch, E.F., Sambrook, J.: Molecular cloning: A laboratory manual, Cold Spring Harbor Laboraotory, N.Y. (1982).
20. Undombunditkul, M.: Increased expression of a cloned penicillin acylase gene in <u>Escherichia coli</u> by using a strong promoter, M.Sc. Thesis, Mahidol University, Bangkok, Thailand (1989).
21. Guyer, M.S.: J. Mol. Biol. <u>126</u>, 347 (1978).

THE ENZYME AND GENE OF THE THERMOSTABLE PEROXIDASE FROM BACILLUS STEAROTHERMOPHILUS

Suvit Loprasert, Seiji Negoro, Itaru Urabe and Hirosuke Okada

Department of Fermentation Technology
Faculty of Engineering, Osaka University
Suita-Shi, Osaka 565, Japan

Peroxidases are important enzymes; widely used in the clinical field for colorimetric measurement of biological materials. Peroxidases catalyse the oxidation of a large number of aromatic compounds such as phenol, hydroquinoid amines, especially benzidine derivatives. The longest known and best studied peroxidase is horseradish peroxidase but bacterial peroxidases have attracted comparatively little attention.

So far, microbial peroxidases from Pseudomonas fluorescens (cytochrome c peroxidase), Streptococcus faecalis (NADH peroxidase), Escherichia coli (o-dianisidine peroxidase, HP-I), Halobacterium halabium, Rhosopseudomonas capsulata, Pellicularia filamentosa, and Saccharomyces cerevisiae (cytochrome c peroxidase) have been isolated, but there is little information about peroxidase from thermophilic important information about thermal resistance in catalase -peroxidase and will be suitable for practical purposes. In the first part, the thermostable peroxidase from a thermophilic bacterium, Bacillus stearothermo philus, was purified and some of its characteristics were studied. The second part described cloning of the thermostable peroxidase gene including the nucleotide sequence and analysis. The peroxidase gene of B. stearothermophilus was manipulated to be overexpressed in E. coli and the simple method of enzyme purification was developed and described in the last part.

Table 1. Purification of peroxidase from B. stearothermophilus

Step	Total protein mg	Total activity unit	Specific activity unit/mg	Yield %
Crude extract	4100	610	0.15	100
Ammonium sulfate	2700	480	0.18	79
DEAE-Sephadex A-50 (I)	260	440	1.6	72
DEAE-Sephadex A-50(II)	170	400	2.3	66
Phenyl-Sepharose CL-4B	40	360	9.9	59
Sephadex G-200	7	120	17.1	20
HPLC-DEAE-5PW	3.3	58	17.7	10

Biotechnology and Environmental Science: Molecular Approaches
Edited by S. Mongkolsuk et al., Plenum Press, New York, 1992

201

Table 2. Substrate specificity of peroxidase

Substrate	Relative activity(%)
1. Dichlorophenol +Aminoantipyrine	100
2. Phenol+Aminoantipyrine	37
3. o-Aminophenol	1.6
4. Pyrogallol	1.3
5. Catechol	0
6. Guiacol	0

For substrate 1 and 2, conditions were as described in the Methods. For substrates 3 to 6, the concentration of substrate in the reaction mixture was 0.7 mM; absorbance measured at 500 nm for 1 and 2, 480 nm for 3 and 5, 420 nm for 4, and 430 nm for 6. Increases in 1 min were compared and expressed as relative activity. For dichlorophenol+aminoantipyrine 100% relative activity was 18.6 mmol min $^{-1}$ ml^{-1}.

PURIFICATION AND CHARACTERIZATION OF THE THERMOSTABLE PEROXIDASE FROM B. STEAROTHERMOPHILUS IAM 11001

A peroxidase from B. stearothermophilus was purified to homogeneity by using a series of column chromatography (table 1). The enzyme (Mr 175,000) was composed of two subunits of equal size, and showed a Soret band at 406 nm. On reduction with sodium dithionite, absorption at 434 nm and 558 nm was observed. The spectrum of reduced pyridine hemochrome showed peaks at 418, 526 and 557 nm; the reduced minus oxidized spectrum of pyridine hemochrome revealed peaks of 418, 524 and 556 nm with a trough at 452 nm. These results indicate that the enzyme contained protoheme IX as a prosthetic group. The optimum pH was 6 and the apparent optimum temperature was 70°C. The enzyme was relatively stable up to 70°C; at 30°C it was stable for a month. The enzyme had peroxidase activity toward a mixture of 2,4 -dichlorophenol and 4-aminoantipyrine as described in table 2 with a K_m for H_2O_2 of 1.3 mM and a V_{max} was 8.7×10^3 min^{-1}. It also acted as catalase with a K_m for H_2O_2 of 7.5 mM and a V_{max} of 2.3×10^5 min^{-1}.

CLONING AND SEQUENCING OF THE PEROXIDASE GENE FROM B. STEAROTHER-MOPHILUS

The gene encoding a thermostable peroxidase was cloned from the chromosomal DNA of B. stearothermophilus IAM11001 in E. coli. The nucleotide sequence of the 3.1-kb EcoRI fragment containing the peroxidase gene (perA) and its flanking region was determined. A 2193-base-pair open reading frame encoding peroxidase of 731-amino-acid residues (Mr 82,963) was observed(Fig.1). A Shine-Dalgarno sequence was found 9 bp upstream of the translational starting site. The deduced amino acid sequence coincides with those of the amino terminal and four peptides derived from the purified peroxidase of B. stearothermophilus IAM11001. E. coli harboring a recombinant plasmid containing preA produces a large amount of thermostable peroxidase which comigrates on polyacrylamide gel electrophoresis with the B. stearothermophilus peroxidase. Peroxidase of B. stearothermophilus has 48% overall homology in the amino acid sequence to catalase-peroxidase of E. coli. There are three

```
B. st.   M----ENQNRQNAAQCPFHESVTNQS-SNRTTNKDWWPNQLNLSILHQHDRKTNPHDEEFNYAEEFQKLDYWALKEDLR
          *   * *     *****    **    **  *******   * **    **  * * *  ** ****  ** **
E. coli  MSTSDDIHNTTATGKCPFHQGGHDQSAGAGTTTRDWWPNQLRVDLLNQHSNRSNPLGEDFDYRKEFSKLDYYGLKKDLK

         KLMYESQDWWPADYGHYGPLFIRMAWHSAGTYRIGDGRGGASTGTQRFAPLNSWPDNANLDKARRCYGRSKRNTGTK-S
          *  **** ***  * * * ********** *****  ****  ********  ******  *  *    *   *
         ALLTESQPWWPADWGSYAGLFIRMAWHGAGTYRSIDGRGGAGRGQQRFAPLNSWPDNVSLDKARRLLWPIKQKYGQKIS

         LGPICSFWRAMSLLNRWVEKRLDSAAGPLTSGIRKKTFIGDRK-KSGSPLNAIPVIASSKTRSPRANGVNLRQPRRAGR
          **  *   *  *
         WADLFILAGNVALENSGFRTFGFGAGREDVWEPDLDVNWGDEKAWLTHRHPEALAKACLGATEMGLIYVNPEGPDHSG-

         QAGSKSRGISAETFRRMGMNDEETVALIAGGHTFGKAHRGGPATHVGPEPEAAPIEAQGLGWISSYGKGKGSDTITSGI
          *       *   * * ***  ********************** **  *   *** ******* *****  * ****
         EPLSAAAAIRA-TFGNMGMNDEETVALIAGGHTLGKTHGAGPTSNVGPDPEAAPIEEQGLGWASTYGSGVGADAITSGL

         EGAWTPTPTQWDTSYFDMLFGYDWWLTKSPAGAWQWMAVDPDEKDLAPDAEDPSKKVPTMMMTTDLALRFDPEYEKIAR
          * **  *****      *  ** *  *****  * ***   *     ** *****   * *** ****** *** *
         EVVWTQTPTQWSNYFFENLFKYEWVQTRSPAGAIQFEAVDAPE--IIPDPFDPSKKRKPTMLVTDLTLRFDPEFEKISR

         RFHQNPEEFAEAFARAWFKLTHRDMGPKTRYLGPEVPKEDFIWQDPIPEVDYELTEAEIEEIKAKILNSGLTVSELVKT
          ** * *  *************************** *****  *  * * *   ** *    *** *
         RFLNDPQAFNEAFARAWFKLTHRDMGPKSRYIGPEVPKEDLIWQDPLPQPIYNPTEQDIIDLKFAIADSGLSVSELVSV

         AWASA--ARSATRISAATNGRRIRLAPQKDWEVNEPERLAKVLSVLRGHPARTAEKSKHRRLDRLGGTLRWKRQPATPA
          *****        * * *  *** ** *           *                *      *
         AWASASTFRGGDKRGGA-NGARLALMPQRDWDVNAAAVRA--LLVLEKIQKESGKASLADIIVLAGVVGVEKAASA-AG

         LMSKCHF---SLAAAMRHKSKPMSKALPCWNRSQMASATIKSKSTRFRRKSCSSTKPSSSADRPRNDGLSWRFARVGPN
          *    *          *     *  * *    *  *  ** *  *      * *      * * *    *   * *
         LSIHVPFAPGRVDARQDQTDIEMFELLEPIADGFRNYRARLDVSTTESLLIDKAQQLTLTA--PEMTALVGGMRVLGAN

         YRHLPHGVFTDRIGVLTNDFFVNLLDMNYEWVPTD--SGIYEIRDRKTGEVRWTATRVDLIFGSNSILRSYAEFYAQDD
          ******  *** ********** *  * *** ***    *** **   * *** *  ******  *    *
         FDGSKNGVFTDRVGVLSNDFFVNLLDMRYEWKATDESKELFEGRDRETGEVKFTASRADLVFGSNSVLRAVAEVYASSD

         NQEKFVRDFINAWVKVMNADRFDLVKKARESVTA*
          **** **  *******  ****
         AHEKFVKDFVAAWVKVMNLDRFDL----------L
```

Figure 1. Comparison of amino acids sequence between <u>B. stearothermophilus</u> and <u>E. coli</u> hydroperoxidase.

regions in the sequence where the homology are 70-80%. The positions of distal and proximal histidine are also conserved among many peroxidases.

OVERPRODUCTION AND SINGLE STEP PURIFICATION OF THE <u>B. STEAROTHERMO-PHILUS</u> PEROXIDASE IN <u>E. COLI</u>

The cloned peroxidase gene from <u>B. stearothermophilus</u> was highly expressed in <u>E. coli</u>. Using the high copy number plasmid which is temperature-sensitive and its own strong promoter, the thermostable peroxidase was produced at 28% of the total cell proteins when the cells were grown at 42°C. The enzyme could be easily purified from <u>E. coli</u> by heat treatment (70°C 10 min) and single column Sephadex G-200 chromatography. From a 200 ml culture, 30 mg of purified enzyme was obtained which is 270 folds higher yield than isolation from the parent organism (<u>B. stearothermophilus</u>). The peroxidase produced by <u>E. coli</u> showed a thermostability, heme type and content identical with those of the peroxidase produced by <u>B. stearothermophilus</u>.

This peroxidase is interesting in terms of its dual activity of peroxidase and catalase as well as its thermostability and substrate specificity. The attempts to engineer this enzyme are being made via a site-directed mutagenesis in order to understand the molecular mechanism of this thermostable peroxidase.

REFERENCES

Loprasert, S., S. Negoro, and H. Okada. 1988. Thermostable peroxidase from Bacillus stearothermophilus J. Gen. Microbiol. 134: 1971-1976.

Loprasert, S., S. Negoro, and H. Okada. 1989. Cloning, nucleotide sequence, and expression in Escherichia coli of the Bacillus stearothermophilus peroxidase gene (perA). J. Bacteriol. 171: 4871-4875.

Loprasert, S., I. Urabe, and H. Okada. 1989. Overproduction and single-step purification of Bacillus stearothermophilus peroxidase in Escherichia coli. Appl. Microbiol. Biotechnol. 32: 690-692.

Triggs-Raine, B. L., B. W. Doble, M. R. Mulvey, P. A. Sorby, and P. C. Loewen. 1988 Nucleotide sequence of Kat G, encoding catalase HP I of Escherichia coli. J. Bacteriol. 170: 4415-4419.

APPLICATIONS OF A MULTICOPY PHENOTYPE APPROACH

J.E. Trempy

Department of Microbiology
Oregon State University, Corvallis, OR

INTRODUCTION

Traditional approaches to the identification of regulatory proteins in a particular cellular pathway have revealed an impressive array of controls for gene expression. Typically these approaches involved either biochemical fractionation and assay of cell extracts or a genetic selection based on an increase or decrease in a defined activity. Additionally, genetic complementation for a mutation has been used to identify gene functions for a variety of pathways and led to the isolation of those genes. Despite such successes, these approaches have shortcomings when applied to the identification of regulatory functions. For example, the regulatory protein of interest might not exist in quantities large enough for detection, and a suitable activity assay might not be available. Likewise, a selection for a mutation in the regulatory function of interest might give rise to a lethal phenotype, thus defeating the goal of the selection. Hartwell and coworkers present several limitations of a conditional lethal mutation approach to support the suggestion that the identification of approximately 50 cell division cycle gene products through mutation falls short of the predicted 400 gene products required in *Saccharomyces cerevisiae* mitotic chromosome replication and segregation (1). They indicate that an approach based on a conditional lethal selection will identify those activities that are : 1) involved in morphological changes, 2) not present as functionally redundant genes, 3) involved in a essential function instead of a fidelity function, or 4) represent nonrandom incidence of temperature sensitive mutations (2). Thus, classes of regulators whose activities do not meet these criteria would be missed.

Recently, several articles have appeared which described an attempt to circumvent the restrictions of a genetic or biochemical approach. This approach draws from DNA recombinant technology and is based on the premise that a regulatory activity can be selected from among a multicopy plasmid library by suppression or enhancement of a particular cellular phenotype. The logic behind a multicopy phenotype approach differs somewhat from a genetic complementation approach; in the latter a known missing function (e.g. auxotrophy) is replaced by the introduction of a homologous function. Hartwell and coworkers have identified two new genes that function in yeast mitotic chromosome transmission by screening for activities which when overproduced from high copy number plasmid vectors, conferred a high frequency of chromosome loss or recombination (2). They successfully reasoned that gene products involved in the fidelity of mitotic chromosome transmission could be identified by creating an abnormal stoichiometry of gene products required in precise amounts, such as those found in multicomponent structures (2, 3). Characterization of these two regulatory genes, MIF1 and MIF2 (mitotic fidelity), revealed that the single copy gene function parallels the multicopy function for which they were selecting. A deletion in the MIF1 locus reduced the fidelity of chromosome transmission and a deletion in the MIF2 locus resulted in stage specific cell cycle arrest. These results indicated that the multicopy phenotype approach used for the selection of these activities yielded data which was not derived from an unnatural situation in the cell (2).

Sin: AN ACTIVITY WITH DIAMETRICAL FUNCTIONS

Many developmental genes, when present in high copy number, inhibit the stress induced event of sporulation in *Bacillus subtilis*. Smith and coworkers used this information in identififying a new regulatory gene product of cell growth and development with a multicopy phenotype approach. They screened a high copy number plasmid library for activities that inhibited sporulation phenotypically in a sporulation-proficient strain, identifying three *Spo⁻* colonies that contained an identical 4.0 kilobase insert (4). Deletion analysis revealed that this activity was contained within a 1.0 kilobase fragment and sequence analysis revealed that the predicted amino acid sequence of the protein from this open reading frame had a DNA binding domain similar to that found in several regulatory proteins (4, 5). One of the more interesting features of the Sin protein (sporulation inhibitor) is its capacity for multifunctional regulation. When Sin activity was produced from a multicopy plasmid, both sporulation and the production of extracellular proteases associated with stationary phase growth were inhibited, indicating a repressor function (4, 5). However, a deletion in the Sin locus was phenotypically pleiotropic: cells did not become component or motile suggesting a role as an activator of these two cellular functions (4, 5). Smith and coworkers presented strong evidence for the repressing activity of Sin protein using integrated *lacZ* gene fusions to identify the targets of Sin repression, in addition to identifying DNA binding sites for Sin protein through DNA footprinting (4, 5). However, the method by which Sin activates late-growth processes, such as competence and motility, remains to be determined. Smith and coworkers have suggested that Sin functions in developmental switching by virtue of its dual regulatory nature (4, 5).

Alp: AN ACTIVITY, WHEN OVER PRODUCED, REGULATES TWO DISTINCT STRESS RESPONSES

Searching for new energy dependent proteases (other than Lon protease) in *Escherichia coli* led to the identification of Alp activity (activator of Lon-like protease) using a multicopy phenotype approach (6). *alp* cloned on a high copy number plasmid suppressed the phenotypes of sensitivity to DNA damaging agents (e.g. ultraviolet light, methylmethane sulfonate, mitomycin C) and overproduction of capsular polysaccharide in *lon⁻* mutants (6). The suppression of these *lon⁻* phenotypes was a reflection of the increased proteolysis of specific Lon substrates when Alp was overproduced. Although overproducing Alp permitted cells to respond to these two environmental stresses as if Lon protease were present, the mechanism by which Lon and Alp act is quite different. Lon protease directly affects the stability of key regulators in these two stress responses (reviewed in 7). In the case of recovery from DNA damage, in vivo and in vitro data have demonstrated that Lon protease, in the presence of ATP, is directly responsible for the degradation of the cell division inhibitor, SulA (6, 7, 8). Treatment of *lon⁺* cells with a DNA damaging agent gives rise to cells that form short filaments which are resolved into individual cells when normal cell division resumes(9). The recovery from DNA damage correlates with the degradation of SulA (7, 8). Treating *lon⁻* cells in a similar manner gives rise to cells that form long, nonseptated filaments due to the stabilization of SulA; these cells fail to recover and eventually die (7, 8, 9). In the case of capsular polysaccharide production, Lon protease affects the regulation of capsule synthesis by directly affecting the stability of the capsule transcriptional activator, RcsA: cells containing active Lon protease do not overproduce capsule (RscA is degraded) whereas cells mutant for Lon protease over-produce capsule (RcsA is stable) (reviewed in 7). However, these two Lon substrates are not completely stabilized in a *lon⁻* strain (10, 11). The selection of Alp activity through a multicopy phenotype approach was deemed possible because an overexpressed activity could be identified that permitted cells to respond as if Lon was available (6). The cellular phenotypes resulting from overproducing Alp are dependent on a DNA fragment of 750 basepairs, containing a 70 amino acid open reading frame which codes for a highly basic polypeptide (J.E. Trempy and S. Gottesman, manuscript in prep.). It seems likely that this Alp protein is not acting alone to affect the stability of SulA and RscA given its relatively small size when compared to other known proteases. Tn10 mutagenesis was used to identify chromosomal second site suppressors of multicopy Alp activity. *slp* mutations (substitute Lon-like protease) define additional chromosomal components of the Alp system needed for the suppression of the *lon* phenotypes (J.E. Trempy and S. Gottesman, manuscript in prep.). *slp* mutations mapped close to the chromosomal location of *alp* (greater than 90% linkage) but distinct from it (J.E. Trempy and S. Gottesman, manuscript in prep.). A λMu*lac* transcriptional fusion into *slp* has been constructed; expression of *lac* from this fusion increased 6-8 fold when Alp was overproduced in the cell (J.E. Kirby and S. Gottesman, pers. comm.). Therefore, it seems likely that multicopy Alp acts as a positive transcriptional regulator of the *slp* gene(s). *slp* and *alp* genes appear to be arranged in an operon, thus permitting coregulation of a regulatory protein (Alp) and its target (Slp) (J.E. Trempy and S. Gottesman, manuscript in prep.). Sequencing this region and may reveal if *slp* codes for components of a protease that overlaps in substrate specificity with Lon and when activated by Alp has a role in regulating recovery from environmental stress.

MULTICOPY ACTIVITIES FROM *Bacillus subtilis* REGULATE *E. coli* STRESS RE-SPONSE PATHWAY

A multicopy phenotype approach similar to the one described for Alp identification in *E. coli* has been used to identify regulators that participate in the recovery from environmental stresses in *Bacillus subtilis*. Exploiting the two notable phenotypes of *E. coli lon⁻* cells, (overproduction of capsule and sensitivity to DNA damaging agents) *B. subtilis* activities have been identified which, when overproduced from a high copy number plasmid, suppress both defects in *E. coli* (J.E. Trempy, unpublished result). Examiniation of the DNA restriction pattern of plasmids isolated from five positive clones revealed that three unrelated *B. subtilis* inserts coded for activities which, when overproduced, suppress overproduction of capsule in *E. coli lon⁻* cells (D.D. Chen and J.E. Trempy, manuscript in prep.).These three clones also conferred resistance to the DNA damaging agents ultraviolet light, mitomycin C, and methylmethane sulfonate. Capsule production was measured using a capsule gene-β-galactosidase gene (*cps::lac*) transcriptional fusion, thus permitting capsule synthesis to be measured a s a function of β-galactosidase levels. *E. coli lon* cells carrying one of the three *B. subtilis* multicopy activities have β-galactosidase levels similar to those seen in *E. coli lon+ cells* (D.D. Chen and J.E. Trempy, manuscript in prep.). The levels of β-galactosidase expressed from the *cps::lac* fusion corresponds to the phenotypic observation that *E. coli lon* cells are nonmucoid in the presence of one of the three *B. subtilis* multicopy activities, indicating that the fusion itself is not the target. How might suppression of capsule production and sensitivity to DNA damaging agents be occurring? Youngman and coworkers have found that many segments of *B. subtilis* chromosomal DNA cloned into multicopy plasmids such as pBR322 are toxic to *E. coli* (12). In light of this observation, it seems likely these *B. subtilis* activities represent functions compatible with the *E. coli* capsule pathway and its SOS (response to DNA damage) pathway. The response to DNA damaging agents and the production of capsular polysaccharide do not have any other known regulatory features in common except for regulation by energy dependent proteolysis. In addition, preliminary data from experiments using *lacZ* transcriptional fusions to the two Lon substrates, SulA and RcsA, indicate a normal level of SulA and RcsA synthesis from the chromosome in the presence of the *B. subtilis* activities (J.E. Trempy, unpublished result). This eliminates the possibility that the three identified *B. subtilis* activities act as repressors of SulA or RcsA synthesis. Therefore, the three *B. subtilis* activities might code for an energy dependent intracellular protease similar to Lon. Probing a *B. subtilis* whole cell extract with antibody directed against *E. coli* Lon protein detected a homologous protein (13). These *B. subtilis* activities may represent proteases that can substitute for Lon protease either because they function similarly to Lon or because of an overlap in substrate specificities. Additionally, if the only mechanism in common regulating these two emergency response pathways is proteolysis, then the *B. subtilis* activities may not be the protease itself. These activities could represent a positive regulatory protein capable of correctly activating an *E. coli* protease which has overlapping substrate specificities with Lon. Another possibility is that the *B. subtilis* activities may be tagging the *E. coli* Lon substrates for proteolysis and this modification can be interpreted by *E. coli* proteolytic machinery. Experiments are in progress to elucidate the mechanism by which the *B. subtilis* activities function to regulate two emergency responses in *E. coli* and to determine their natural role in *B. subtilis*.

CONCLUSIONS

A multicopy phenotype approach to identify new regulatory proteins is not without drawbacks. Overexpression of an activity is not a "real life" situation for the cell. Hartwell and coworkers remark that this method, even while conveniently providing the cloned wildtype gene for further analysis, might fool the investigator; overexpression of any activity might give the desired phenotype, or lead to cell death (2). However, there are ways to circumvent these problems and the investigator need only realize that, when using a multicopy phenotype approach, the next obvious step after the identification of an activity will be to determine its role, as a single copy function, in the cell.

ACKNOWLEDGEMENTS

I thank William Brusilow, Donald Chen and Jim Marks for helpful discussions and for critical review of this manuscript. I thank Susan Gottesman for both the technical and intellectual training I received while I was a member of her laboratory.

LITERATURE

1. Pringle, J.R. and L.H. Hartwell. 1981. The *Saccharomyces cerevisiae* cell cycle. In: The Molecular Biology of the Yeast Saccharomyces, J. Strather, E. Jones, and J.R. Broach, eds. Cold Spring Harbor, NY: Cold Spring Harbor Laboratory. pp. 97-142.

2. Meeks-Wagner, D., J.S. Wood, B. Garvik, and L.H. Hartwell. 1986. Isolation of two genes that affect mitotic chromosome transmission in *S. cerevisiae*. Cell 44: 53-63.

3. Meeks-Wagner, D. and L.H. Hartwell 1986. Normal stoichiometry of histone dimer sets is necessary for high fidelity of mitotic chromosome transmission. Cell 44: 43-52.

4. Gaur, N.K., E. Dubnau, and I. Smith. 1986. Characterization of a cloned *Bacillus subtilis* gene that inhibits sporulation in multiple copies. J. Bacteriology 168: 860-869.

5. Smith, I. 1989. Initiation of sporulation. In: Regulation of procaryotic development. I. Smith, R.A. Slepecky, and P. Setlow. Washington, D.C.: American Society for Microbiology. pp. 185-210.

6. Trempy, J.E. and S. Gottesman. 1989. Alp, a suppressor of Lon protease mutants in *Escherichia coli*. J. Bacteriology 171: 3348-3353.

7. Gottesman, S. 1989. Genetics of Proteolysis in *Escherichia coli*. Annual Review of Genetics 23: 163-168.

8. Mizusawa, S. and S. Gottesman. 1983. Protein degradation in *Escherichia coli*: the *lon* gene controls the stability of SulA protein. Proc. Natl. Acad. Sci. USA 80:358-362.

9. Howard-Flanders, P., E. Simon, L. Theriot. 1964. Genetics 49: 237-246.

10. Maurizi, M.R., P. Trisler, and S. Gottesman. 1985. Insertional mutagenesis of the *lon* gene in *Escherichia coli*: lon is dispensable J. Bacteriol. 164: 1124-1135.

11. Torres-Cabassa, A.S. and S. Gottesman. 1987. Capsule synthesis in *Escherichia coli* K12 is regulated by proteolysis. J. Bacteriol. 169:981-989.

12. Youngman, P., J.B. Perkins, and R. Losick. 1984. A novel method for the rapid cloning in *Escherichia coli* of *Bacillus subtilis* chromosomal DNA adjacent to Tn917 insertions. Mol. Gen. Genet. 195: 424-433.

13. Arnosti, D.N., V.L. Singer, and M.J. Chamberlin. 1986. Characterization of heat shock in *Bacillus subtilis*. J. Bacteriol. 168: 1243-1249.

ANALYSIS OF EPITOPES IN THE CHOLERA FAMILY OF ENTEROTOXINS

Mohammad Kazemi[1] and Richard A.Finkelstein

Department of Molecular Microbiology and Immunology
School of Medicine, University of Missouri-Columbia
Columbia, MO 65212

ABSTRACT

It is now over 30 years since the cholera enterotoxin was discovered in India and more than 20 years since it was first purified to homogeneity. Early expectations that, as with diphtheria and tetanus, a toxoid vaccine would solve the cholera problem have not been fulfilled. We are examining the reasons why cholera toxin (CT) vaccines have failed to provide effective immunity.

Studies with polyclonal (pAbs) and monoclonal (mAbs) antibodies reveal significant differences among the CTs and CT-related enterotoxins (LTs). All field trials of toxoid vaccines have used CT-1 antigen against the prevailing epitype of CT, usually CT-2. Single amino acid substitutions in the immunodominant B-subunit protein have profound effects on immunological reactivity. We have employed pAbs and mAbs, genetically engineered chimeric B-subunit proteins, checkerboard immunoblotting (CBIB), synthetic peptides and their antisera, and sequential overlapping synthetic hexapeptides representing the B-subunit chain to identify epitopes in the CT family. Reactivity of some, but not all, synthetic hexapeptides with epitope activity with polyclonal antisera can be reduced by pretreatment of the serum with the native toxin protein indicating that certain continuous epitopes are exposed on the surface of the native protein. Interestingly, the holotoxin was more effective in this regard than the native B-subunit pentamer indicating a conformational difference between the two. Human convalescent sera gave diffuse patterns of reactivity with discontinuous (conformational) epitopes. A major tetrapeptide epitope has been identified in a conserved region of the protein. Substitution of amino acids in this region can be permissive or can reduce or eliminate activity. Studies like these may lead to the development of rational synthetic peptide vaccines, but a vaccine composed only of CT antigen can only protect against severe life-threatening diarrhea, not all the diarrhea caused by cholera vibrios.

---

[1] Present address : Hyland Division, Baxter Healthcare Cerp., 1720 Flower Ave., Duarte, CA 91010

Biotechnology and Environmental Science: Molecular Approaches
Edited by S. Mongkolsuk et al., Plenum Press, New York, 1992

INTRODUCTION

The cholera enterotoxin (CT) which is responsible for the severe, life-threatening diarrhea of cholera was discovered in 1959 (De, 1959, Dutta *et al.*, 1959) and was purified to homogeneity 10 years later (Finkelstein and LoSpalluto,1969). Subsequently, it was recognized that a variety of enteric bacteria produce diarrheagenic enterotoxins that are structurally, functionally, and immunologically related to CT (reviewed in Finkelstein,1988). These include, particularly, several heat-labile enterotoxins (LTs) produced by *Escherichia coli* strains of human (H) origin and an LT produced by *E.coli* strains of porcine (P) origin. Each of the CT-related enterotoxins consists of a 11.5kDa B-subunit homopentamer (~57kDa), which is responsible for binding of the enterotoxin to a glycolipid, the G_{M1} ganglioside, on the target cell membrane, and a 28 kDa A-subunit, whose A1 peptide enzymatically ADP-ribosylates the GTP-binding protein which regulates adenylate cyclase. The resulting elevated level of cAMP causes hypersecretion of electrolytes and water.

Despite the fact that cholera vaccines have been used since the late 1800s, an effective and economical vaccine which provides long lasting immunity against cholera without side effects has yet to be developed and deployed. It is known that the disease itself provides lasting immunity but the involvement of anti-toxic immunity in that process is yet to be established. Early expectations that a toxoid vaccine would provide such immunity against cholera have not been fulfilled in part because the earlier studies used chemically inactivated toxoids with reduced antigenicity; in part because toxin antigen CT-1 was used against *Vibrio cholerae* strains which produced CT-2; and also because *V. cholerae* strains have, in addition to CT, as-yet-undefined mechanisms for causing milder diarrheal disease. We are attempting to define further the reasons why cholera toxin vaccines have failed to provide effective immunity. In the holotoxins the B-subunit protein is immunodominant: anti-B antibodies predominate in hyperimmune antitoxin sera and in the serum of cholera convalescents, and the relatedness of the CT-related enterotoxin family is primarily though their B-subunit proteins. Our studies, in part as reported here, involve the immunological characterization of the B-subunit of cholera toxin at the molecular level to help understand the potential of toxin immunity. The ultimate goal of this study is to evaluate the possibility of developing a peptide vaccine that can immunologically prevent the receptor-binding process and which might provide some degree of immunity against cholera and against the CT-related toxin-mediated diarrheas which are far more prevalent globally than is cholera itself. In this report, we have examined immunological cross-reactivities of various CT-related antigens using a new technique, checkerboard immunoblotting (CBIB; Kazemi and Finkelstein, 1990a). We have also scanned the CT-B molecule to localize its potential epitopic regions by utilizing the synthetic peptides-on-pins approach of Geysen *et al.*(1987).

IMMUNOLOGICAL CROSS-REACTIVITY STUDIES

Checkerboard immunoblotting (CBIB), developed in our laboratory (Kazemi and Finkelstein,1990a), facilitates the examination of the immunological cross-reactivities of multiple antigen/antibody combinations. CBIB is performed in an apparatus, such as the Miniblotter 45 (Immunetics, Cambridge, MA.), consisting of two acrylic plates in which the top plate contains parallel grooves in its inner surface which form parallel channels on top of the membrane on which reagents are incubated. In the CBIB assay, antigens are immobilized on nitrocellulose membranes in the form of parallel lanes which are then reacted with antibodies in similarly applied parallel lanes perpendicular to the antigen lanes. After the

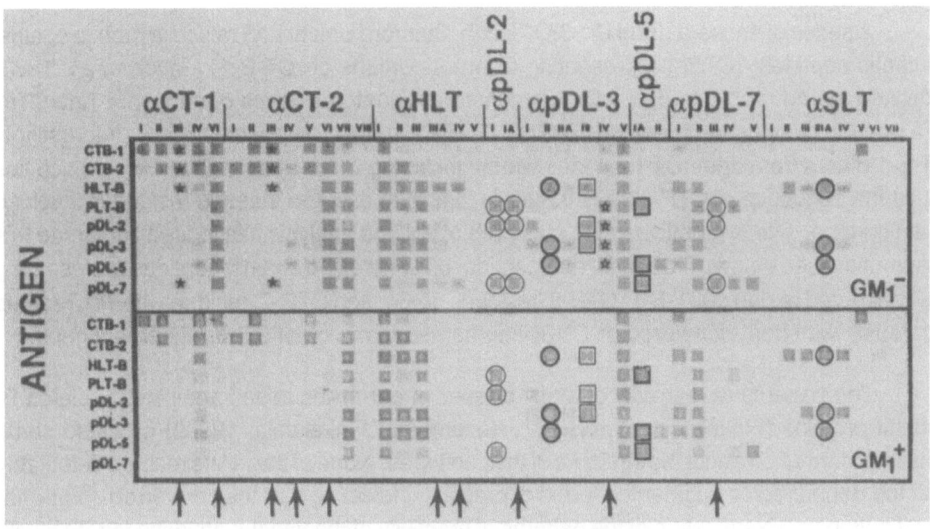

MONOCLONAL ANTIBODY

Figure 1. CBIB assay of the reactivities of various mAbs with untreated (G$_{M1}$⁻ panel) as well as G$_{M1}$ ganglioside treated (G$_{M1}$⁺ panel) CT-B-related antigens. Reactions which were blocked by G$_{M1}$ treatment are shown by arrows. Complementary-type epitopes (see the text) are also identified (☆vs★; ○vs○ and □vs□).

development of the blots with appropriate secondary antibodies and substrate systems, positive reactions appear as small colored squares at the intersection of antibodies and antigens.

We applied CBIB in studying cross-reactivities of a variety of purified CT-related B-subunit protein antigens such as CT-B-1 and CT-B-2 (from classical Inaba 569B and El Tor Ogawa 3083 strains of *V. cholerae*, respectively) and H-LT-B-1 and P-LT-B (from human and porcine strains of *E.coli*, respectively). We also used four genetically engineered chimeric P-LT-Bs (pDLs2,3,5,and 7) in which a single or two amino acids were substituted from the corresponding H-LT-B-1 residues (described in Finkelstein *et al.*, 1987). Antigens were used in their native, as well as in variously denatured or CNBr-fragmented forms, and reacted with various polyclonal and monoclonal antibodies and with several convalescent human sera.

Polyclonal hyperimmune antisera raised against various CT-related immunogens, as reported previously (Kazemi and Finkelstein, 1990b), exhibited various degrees of cross-reactivities. The homologous reactions were strongest indicating that, regardless of their overall immunological similarities, each protein also contains its own unique antigenic determinants. Some antibodies reacted more strongly with the denatured or fragmented proteins while others were only reactive with the native form. This variation in the immune response to a single immunogen was evident even between sera from animals within the same species. Our mouse antisera raised against eight CT-related proteins did not recognize the denatured or fragmented forms of the antigens. Apparently the immune response of these animals was primarily directed toward conformational type epitopes.

A series of antisera (from Dr. C.O. Jacob, Stanford University) raised in rabbits against synthetic peptides (CTPs) representing different portions of CT-B-1 (Jacob *et al.*, 1983) exhibited strong reactions only with fragmented or denatured forms of antigen. Anti-CTP-1 (a peptide composed of residues 8-20 of CT-B-1) reacted only with CT-B-1. Interestingly CT-B-1 differs from other CT-related antigens (including CT-B-2) in residue His[18], which lies within the sequence of CTP-1. It is therefore logical to assume that this residue is involved in the epitope. Our synthetic peptide approach of epitope analysis has indeed confirmed this assumption (see below). Two other anti-peptide antisera raised against peptides representing the conserved region of CT-B-1 (CTP-3, residues 50-64, and CTP-7, residues 45-64), reacted vigorously with heat-denatured or CNBr-fragmented forms of all CT-B-related antigens.

The patterns of reactivity of many classes of our mAbs raised against CT-related B-subunit proteins (Finkelstein *et al.*, 1987, Kazemi and Finkelstein, 1990b) provided some detailed information about the structure of their epitopes. None of the mAbs reacted detectably with the denatured or fragmented forms of proteins, indicating that they preferred conformational epitopes. Examination of the patterns of reactivity of the various antigens together with their amino acid differences disclosed the involvement of certain amino acids in the epitopes

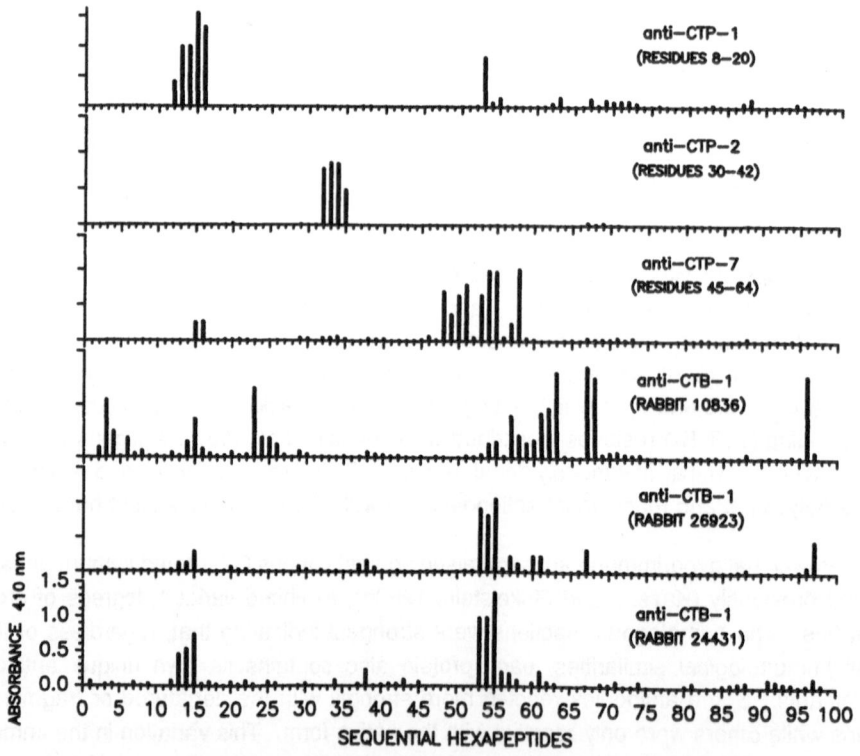

Figure 2. Patterns of reactivity of sequential overlapping hexapeptides-on-pins spanning the CT-B-1 amino acid sequence with αCT-B-1 antisera (bottom three panels) and with antisera raised against CTP peptides representing different portions of CT-B-1 (top three panels).

for a number of these mAbs (reviewed in Finkelstein *et al.*, 1987, Kazemi and Finkelstein, 1990b). For example, as shown in Figure 1, α H-LT class II mAb reacted only with those antigens that possess Ala[46] (except CT-B-1). This suggests Ala[46] in combination with the CT-B-2/LT-specific Tyr[18] (CT-B-1 is the only antigen that has His at position 18) is involved in defining this epitope. Other similar type comparisons indicated that single amino acid substitutions can have profound effects on the structure of many epitopes. Interesting examples are the reactions of α CT-1 class I and α CT-2 class 1 which are very specific for their homologous antigens. One way or another the different amino acids which are specific to these proteins must be involved in the formation of the epitopes recognized by these two mAbs. Another interesting observation was that a single amino acid substitution in an antigen (e.g., in LT-related antigens, changing the Lys[102] of P-LT-B to Glu[102] of H-LT-B) resulted in formation of one epitope (e.g., the one recognized by αpDL-3 class III) and disappearance of another epitope (e.g., the one recognized by αpDL-5 class IA). Since such epitopes seem to appear and disappear reciprocally, we call them "complementary epitopes" (as identified in Fig. 1). In other instances, the mAbs were promiscuous in their reactivity, recognizing common epitopes in the family.

The effects of receptor-binding on the reactivity of mAbs with CT-B-related antigens were examined by saturating the immobilized antigens with G_{M1} ganglioside prior to their reaction with antibodies. As a result of this treatment, the reactivity of several mAbs was abolished (Fig.1) indicating that their epitopes are likely to be located at or near a G_{M1}-binding site. It is also conceivable that the receptor binding process affects these epitopes by altering the native conformation of the protein.

Human convalescent sera reacted primarily with the native form of the antigens emphasizing the role of conformational epitopes in those reactions. Sera from American volunteers challenged with an El Tor strain of *V. cholerae* reacted better with CT-2 than with CT-1. Whereas pretreatment with native CT-2 removed all their reactivity, pretreatment with CT-1 left residual reactivities with CT-2 indicating that these individuals had responded to additional epitopes on CT-2 (results not shown).

EPITOPE SCANNING WITH SYNTHETIC PEPTIDES-ON-PINS

The peptides-on-pins technique, developed by Geysen *et al.*, 1987, allows high resolution scanning of the primary structure of proteins for identification of potential antigenic determinants. Overlapping hexapeptides spanning the entire amino acid sequence of CT-B-1 were manually synthesized on the tips of derivatized plastic pins (from Cambridge Research Biochemicals, Wilmington, DE) with a 96 well plate format. Peptides covalently bound on pins were examined for their reactivities with various antibodies raised against CT-B related antigens in ELISA type assays. From the patterns of reactivities of the peptides with various antibodies, the structures of their epitopes could be predicted. This technique is particularly suitable for studying continuous (sequence-related) epitopes. Covalently-bound peptides-on-pins can be reused many times by disrupting the antibody-peptide bond after each reaction using SDS and sonication.

Results of these studies are presented in detail elsewhere (Kazemi and Finkelstein, 1991). The reliability of the assay was assessed by the reactions of αCTP peptide antisera (Fig. 2, top 3 panels). Anti-CTP antibodies recognized only those hexapeptides that were within the sequence of the original immunizing CTP thus establishing that the synthetic peptides-on-pins contained the expected amino acid sequences.

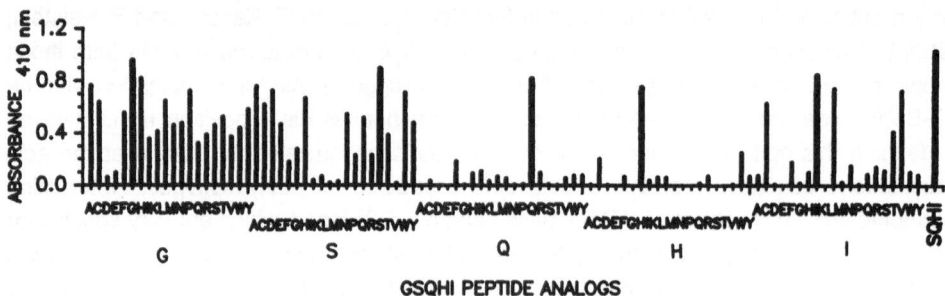

Figure 3. Pattern of reactivity of a CT-B-1 antiserum (rabbit#24431) with amino acid substituted peptide analogs of pentapeptide GSQHI. The bars representing the reactivity of parent peptides are highlighted and the arrow indicates the reactivity of the original SQHI peptide by itself.

Anti-CTP-1 (residues 8-20 of CT-B-1) antiserum reacted with hexapeptides #13-16 (the peptide numbers correspond to the number of their N-terminus residue in the CT-B-1 sequence) suggesting that the epitope is the tripeptide Gln-Ile-His (residues 16-18 of CT-B-1) which is common to these peptides. Similarly αCTP-2 (residues 30-42 of CT-B-1) antiserum identified the tripeptide Arg-Glu-Met (residues 35-37) as a reactive epitope. The αCTP-7 (residues 45-64 of CT-B-1) antiserum identified multiple closely located epitopes in the conserved region of CT-B-1. One of these epitopes was the tetramer Ser-Gln-His-Ile (residues 55-58) represented by the reactivities of hexapeptides 53-55. This epitope was also recognized, although to different degrees, by most αCT-B-related antisera tested, including convalescent human sera. Anti-CT-B-1 sera from rabbits 24431 and 26923 (Fig.2, bottom two panels) reacted strongly with this epitope. Additionally, both of these sera contained antibodies that gave weak reactions with hexapeptides that appeared in the αCTP-1 epitope region. This apparent cross-reactivity evident between CTP-1 and CTP-7, shown in Figure 2, is attributed to the presence of three identical (but in different order) amino acids, Gln[16]-Ile[17]-His[18] (of CTP-1) and Gln[56]-His[57]-Ile[58] (of CTP-7), in these two peptides. Serum from rabbit 10836 (Fig.2) identified additional epitopic regions along the CT-B-1 sequence, of which were many located in the conserved region.

Adsorption of antibodies that were strongly reactive with the SQHI (Ser[55]-Gln[56]-His[57]-Ile[58]) epitope with native CT-related proteins resulted in significant reduction in their reactivities, indicating that this epitope is at least partially exposed on the native proteins. Previous work by Jacob *et al.* (1984) revealed that antibodies raised against peptides representing the conserved region of CT-B-1 (such as CTP-7), and to a lesser degree αCTP-1 antiserum, partially neutralized CT-activity. It was also shown by Jacob *et al.*(1986a and b) that immunization of rabbits with a single dose of CTP-3 synthetic peptide (residues 50-64 of CT-B-1) primed for a vigorous immune response to subsequent administration of a subimmunizing dose of the three CT-related enterotoxins tested. Further analysis of the structure of the epitopes of this region may provide useful information for vaccine development.

The role of each amino acid residue in the reactivity of the SQHI epitope with antibodies was studied by amino acid substitution. All possible analogs of the pentapeptide GSQHI (the SQHI epitope containing the extra N-terminus Gly[54] of the CT-B-1 sequence), in which every residue was replaced with all 20 different amino acids, were synthesized and tested for their reactivities with antibodies. Results of reactions of these analogous peptides with the αCT-B-1 antiserum from rabbit #24431 (Fig.3) indicated that neither the Gln[56] nor the His[57] residues were replaceable with any other amino acid. Such replacements totally abolished the reactivities of the SQHI epitope. Presumably these two residues are essential to specific binding with antibody. Results of this experiment also indicated that Ser[55], and to a lesser degree Ile[58], can be replaced with a few other amino acids with minimal effect on reactivity of the epitope. The extra Gly[54] on the other hand, can be replaced with many other amino acids with little or no effects on the reactivity of the epitope, with the exception of the negatively charged Glu and Asp: these substitutions abolished the reaction. Adsorption of this antiserum with native CT-1 or with native CT-B-1 abolished its reactivity with these peptides which confirms that SQHI is significantly exposed on the surface of the native proteins. The CT (holotoxin) was a more effective adsorbant than the purified B-subunit protein, suggesting that the conformation of the B-subunit protein is different in its purified form from its conformation in the holotoxin.

Identification and characterization of the immunobiologically important peptides of the receptor-binding B-subunit proteins can be useful to the development of peptide vaccines that may contribute to immunity against the life-threatening diarrhea caused by CT-related enterotoxins.

ACKNOWLEDGMENT

This work was supported by Public Health Service grants AI 16776 and AI 17312 from the National Institute of Allergy and Infectious Diseases.

REFERENCES

De,S.N., 1959, Enterotoxicity of bacteria-free culture filtrate of Vibrio cholerae, Nature, 183:1533.

Dutta,N.K.,Panse, M.W., and Kulkarni, P.R.,1959, Role of cholera toxin in experimental cholera, J. Bacteriol., 78:594.

Finkelstein, R.A., and LoSpalluto, J.J., 1969, Pathogenesis of experimental cholera: preparation and isolation of choleragen and choleragenoid, J.Exp. Med., 130:185.

Finkelstein, R.A., Burks, M.F., Zupan, A., Dallas, W.S., Jacob, C.O., and Ludwig, D.S., 1987, Epitopes of the cholera family of enterotoxins, Rev. Infect. Dis., 9:544.

Finkelstein, R.A., 1988, Cholera, the cholera enterotoxins, and the cholera enterotoxin-related enterotoxin family, pp. 85-102. In: Immunochemical and molecular genetic analysis of bacterial pathogens, P. Owen and T.J.Foster (eds.), Elsevier Science Publishers, Amsterdam, The Netherlands.

Geysen, H.M., Rodda, S.J., Mason, T.J., Tribbick, G., and Schoofs, P.G., 1987, Strategies for epitope analysis using peptide synthesis, J. Immunol. Methods, 102:259.

Jacob, C.O., Arnon, R., and Finkelstein, R.A.,1986a, Immunity to heat-labile enterotoxins of porcine and human Escherichia coli strains achieved with synthetic cholera toxin peptides, Infect. Immun., 52:562.

Jacob, C,O., Grossfeld, S., Sela, M., and Arnon, R., 1986b, Priming immune response to cholera toxin induced by synthetic peptides, Eur. J. Immunol., 16:1057.

Jacob, C.O., Sela, M., and Arnon, R., 1983, Antibodies against synthetic peptides of the B-subunit of cholera toxin:cross-reaction and neutralization of the toxin, Proc. Natl. Acad. Sci.. USA, 80:7611.

Jacob, C.O., Sela, M., Pines, M., Hurwitz, S., and Arnon, R., 1984, Both cholera toxin-induced adenylate cyclase activation and cholera toxin biological activity are inhibited by antibodies against related synthetic peptides, *Proc. Natl. Acad. Sci., USA.*, 81:7893.

Kazemi,M., and Finkelstein, R.A., 1990a, Checkerboard immunoblotting (CBIB):an efficient, rapid, and sensitive method of assaying multiple antigen/antibody cross-reactivities, *J. Immunol. Methods*, 128:143.

Kazemi, M., and Finkelstein, R.A., 1990b, Study of epitopes of cholera enterotoxin-related enterotoxins using checkerboard immunoblotting, *Infect. Immun.*, 58:2352.

Kazemi, M., and Finkelstein, R.A., 1991, Mapping epitopic regions of cholera toxin B-subunit protein, *Molec. Immunol.*, 28: 865.

DETECTION OF INDICATOR BACTERIA AND PATHOGENS IN WATER BY POLYM ERASE CHAIN REACTION (PCR) AND GENE PROBE METHODS

R.M. Atlas, A. Bej, M. Mahbubani, R. Steffan, M. Perlin, J. DiCesare* and L. Haff*

Dept. of Biology, Univ. of Louisville, KY USA
*Perkin-Elmer Corp., Norwalk, CT USA

ABSTRACT

Sensitive methods for detecting coliform bacteria, and pathogenic bacteria (Salmonella and Shigella) in environmental waters that do not require culturing of bacteria were developed by using the polymerase chain reaction (PCR) and gene probes. Cells were collected by filtration and DNA was released by freeze-thaw cycling. PCR amplification of region of lacZ gene was used as a target for detection of total coliform bacteria. A region of the lamB gene was the target for detection of Escherichia coli, Salmonella and Shigella. A region of the uid gene was used for detection of E. coli and Shigella. Biotin was incorporated during PCR amplification and the biotinylated DNA was detected by using membrane-bound poly-T-tailed capture probes. As little as 1 fg of total DNA and as few as 1 cell per 100 ml were detectable by the PCR-gene probe method.

INTRODUCTION

Public health measures require microbiological surveillance of environmental water supplies, e.g., coliform monitoring to indicate potential contamination with enteric pathogens and specific detection of pathogens such as Salmonella. The goal is to detect pathogenic microorganisms that can be transmitted via the fecal-oral route with sufficient sensitivity to ensure the safety of the water. The traditional methods for detecting coliform bacteria rely upon culturing on a medium that selectively permits the growth of Gram-negative bacteria and differentially detects lactose-utilizing (b-galactosidase-producing) bacteria, e.g., using Mac-Conkey's, m-Endo, eosin methylene blue (EMB), or brilliant-green-lactose-bile media[1]. Cultural tests require up to 3 days to confirm the presence of fecal coliforms. Recently the Colilert test based upon detection of b-galactosidase producing bacteria for total coliform and b-galactosidase producing bacteria for fecal coliforms, which while it still requires culturing takes less than a day to complete, has been proposed in the United States as equivalent to conventional plating procedures for detecting coliforms[2, 3].

We report here on our development of noncultural, genetically-based procedures for

Biotechnology and Environmental Science: Molecular Approaches
Edited by S. Mongkolsuk et al., Plenum Press, New York, 1992

219

the environmental detection of coliforms and pathogens in environmental and potable waters based upon the recovery of DNA, and amplification of specific diagnostic target nucleotide sequences by using the polymerase chain reaction (PCR) followed by detection of the amplified DNA with gene probes[4-6].

MATERIALS AND METHODS

Cell Collection and DNA Release

Various procedures, including centrifugation and filtration methods, were tested for the ability to recover "all" cells of target organism in 100 mL water samples. Numerous different types of filters were examined as were several types of lysis buffers and physical methods for releasing DNA from bacterial cells and protozoan oocysts. In one method, cells were collected by filtration through 13 mm 0.5 μm pore size teflon filters, the filters were placed into 100μl diethylpyrocarbonate-treated autoclaved distilled water, and DNA was released by 5 freeze-thaw cycles using ethanol-dry ice and 50°C water baths. In these tests serial dilutions of E. coli cells were added to 100 mL water samples. The success of the method was judged by microscopic observation to determine the efficiency of cell lysis and by PCR amplification-gene probe detection to determine the absence of interesting factors.

PCR Amplification

PCR amplification was performed in a Perkin-Elmer Cetus DNA thermal cycler using a GeneAmp PCR reagent kit and Taq DNA polymerase enzyme. The standard PCR reaction was: total reaction volume 100-150 μl; reaction mixture 50 mM KCl, 10 mM Tris-HCl, 1.5 mM MgCl$_2$, 100 μg/ml gelatin, 200 μM of each of the deoxyribonucleotide triphosphates (dATP, dTTP, dCTP, dGTP), 0.25-1.0 μM of each of the primers and 2.5 units of the Taq DNA polymerase (0.5 μl), 1fg-1 μg of genomic DNA. PCR amplification was performed as follows: program 3 min at 95°C for initial denaturation of target DNA followed by 25-45 cycles of 1 min at 94°C (denaturation), 1 min at 50°C (primer annealing), 1 min at 72°C (primer extension). In some cases biotin was incorporated during PCR amplification by adding biotin-11-dUTP (Sigma) and dTTP with a ratio of 1:3, respectively.

A 0.326 kb region of E. coli lacZ gene was amplified by using 24 mer primers ZL-1675 (5'-ATGAAAGCTGGCTACAGGAAGGCC) and ZR-2025(5'GGTTTATGCAGCAACG AGACGTCA). A 0.309 kb segment of the coding region of the lamB gene of E. coli was amplified using 24 mer primers designated BL-4910 (5'-CTGATCGAATGGCTGCCAGG CTCC) and BR-5219 (5'-CAACCAGACGATAGTTATCACGCA). A 0.153 kb E. coli uidR region was amplified by using 22 mer primers URL-301 (5'-TGTTACGTCCTGTAGAAAG CCC) and URR-432 (5'-AAAACTGCCTGGCACAGCAATT). Multiplex PCR amplification was carried out using equal concentrations (0.2-0.5 μM) of all three sets of primers (lacZ, lamB, and uidR).

Gene Probe Detection

The coliform gene probes were: for the 0.153 kb uidR amplified DNA a 40 mer UR-1 (5'CAACCCGTGAAAYCAAAAAA-CTCGACGGCCTGTGGGCATT) oligonucleotide probe, for 0.326 kb lacZ amplified DNA a 50 mer oligonucleotide probe LZ-1 (5'TGACGT CTCGTTGCTGCATAAACCGACTACAC AAATCAGCGATTTCCATT), and for 0.309 kb lamB amplified DNA a 50 mer oligonucleotide probe LB-1 (5'TGCGTGATAAC-TATCGTCTGGTTGATGGCGCA TCGAAAGAC GGCTGGTTG).

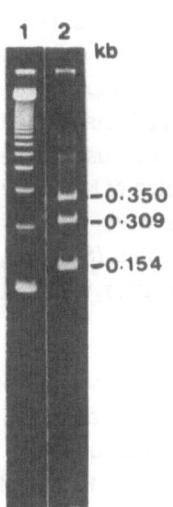

Fig. 1. Examples of specificity of PCR-gene probe detection of bacterial pathogens showing dot blot analysis of the PCR-amplified samples of coliforms using radiolabelled probes. One-tenth (10 ml) of the PCR amplified DNA sample was denatured and spotted on the nylon membrane. Panel A, Total coliforms were amplified and detected by using a probe for the lacZ gene of the E. coli. Panel B, Indicator bacteria E. coli, and other enteric pathogens Salmonella and Shigella were amplified and detected by using a probe for the lamB gene of E. coli. Panel C, Specific detection of indicator bacteria E. coli, and enteric pathogen Shigella using primers and probe for the uidR gene of the E. coli.

In some cases PCR amplified DNAs were detected by ethidium bromide (EtBr) stained polyacrylamide gel electrophoresis (PAGE) and radiolabelled gene probes. In other cases biotin was incorporated during PCR and the biotinylated amplified DNAs were detected using capture probes containing homopolymer dT-tail immobilized on a nylon membrane by UV photofixation[7].

Specificity Determination

To determine the specificity of PCR-gene probe detection of coliform bacteria various coliform and noncoliform bacteria were tested. In total over 100 species were tested including various strains of environmental isolates[5].

Sensitivity Determination

To determine the sensitivity of coliform detection by PCR-gene probes, tests were run using 1 ng - 1 µg E. coli genomic DNA or 1-10 viable cells of E. coli in 100 mL dechlorinated potable water. 100 replicate samples were tested by the PCR-gene probe to determine the distribution of positive signals at low dilutions. Samples were tested by the PCR-gene probe method as described above.

RESULTS

The following bacterial genera showed positive PCR amplification and gene probe detection for lacZ: Citrobacter, Enterobacter, Escherichia, Klebsiella, and Shigella; all other strains tested were negative. Thus, lacZ is a good choice for a genetically based detection system for total coliforms. The following bacteria showed positive PCR amplification and gene probe detection for lamB: Escherichia, Salmonella, and Shigella; all others tested were negative. The following bacteria showed positive PCR amplification and gene probe detection for uidR: Escherichia and Shigella; all others were negative for uidR amplification and gene probe detection. Thus, PCR amplification of lamB and uidR, as demonstrated here, provides a means of monitoring the indicator bacterial species of fecal contamination (E. coli) and also of the enteric bacterial pathogens that cause waterborne disease outbreaks, most importantly Salmonella and Shigella. In the multiplex PCR system, addition of equimolar concentrations of all three primer sets and incorporation of biotin-11-dUTP during PCR amplification of the E. coli genomic DNA resulted in an almost equal quantity of all three target DNAs. Hybridization of the multiplex PCR-amplified DNAs with the immobilized capture probes using Blue-Gene kit (BRL) showed on non-specific signal. By multiplex PCR amplification of lacZ, lamB, and uidR sequences and by immobilized capture probe detection, we were able to detect simultaneously not only the indicator species (E. coli), but also the pathogens (Salmonella and Shigella) whose absence indicates the bacteriological safety of potable water supplies (Fig. 1).

The sensitivity of detection achieved by amplification of lacZ,lamB, and uidR coupled with ^{32}P-labelled gene probes or by colorimetric immobilized capture probe detection assay was equivalent to 1-10 mg of target DNA, i.e., single genome copy (single cell) detection. Tests with dilutions of viable cells of coliform bacteria indicated that 1-5 cells per 100 ml of water could be detected (Fig. 2). Thus, the sensitivity of the PCR-gene probe method is adequate for water quality monitoring to meet U.S. Federal regulations.

Viable cells of E. coli

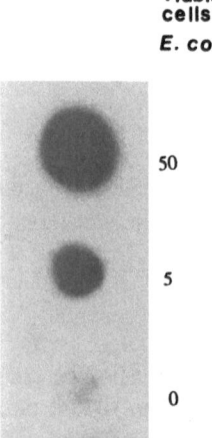

50

5

0

Fig. 2. Example of sensitivity of PCR-gene probe for detecting bacterial indicator in water, showing dot-blot analysis following PCR amplification of serial dilutions of an exponentially growing E. coli culture by using primers for uidR amplification. Viable cells were determined by viable plate count. Control included the sample from the dilution blank.

CONCLUSIONS

In conclusion, both indicator and pathogenic bacteria associated with human fecal contamination of waters could be detected by PCR amplification and gene probes. The use of PCR and gene probes permits both the specificity and sensitivity necessary as a basis for a method for monitoring coliforms as indicators of human fecal contamination of water and also pathogenic bacteria. Cell collection by filtration, DNA release by freeze-thaw cycling, multiplex PCR amplification and nonisotopic gene probe detection techniques, such as immobilized capture probe, can permit a rapid and reliable means of assessing the microbiological safety of waters and should provide an effective alternative methodology to the conventional viable culture methods.

REFERENCES

1. Greenberg, H., Ed., "Standard Methods for the Examination of Water and Wastewater," American Public Health Association, Washington, D.C., 1985.
2. Edberg, S.C. and Edberg, M.M., "A defined Substrate Technology for the Enumeration of Microbial Indicators of Environmental Pollution," Yale Journal of Biological Medicine, **61**, 389-399, 1988.
3. Edberg, S.C., Allen, M.J., Smith, D.B., and the national collaborative study, "National Field Evaluation of a Defined Substrate Method for the Simultaneous Detection of Total Coliforms and Escherichia coli from Drinking Water: Comparison with Presence-absence Techniques," Applied and Environmental Microbiology, **55**, 1003-1008, 1989.
4. Mahbubani, M.H., Bej, A.K., Miller, R., Haff, L., DiCesare, J., and Atlas, R.M., "Detection of Legionella by Using Polymerase Chain Reaction and Gene Probe Methods", Molecular and Cellular Probes, **4**, 175-187, 1990.
5. Atlas, R.M., Bej, A.K., McCarty, S., DiCesare, J., and Haff, L., "Monitoring Microbial Pathogens and Indicator Microorganisms in Water by Using Polymerase Chain Reaction and Gene Probes", Monitoring Water in the 1990's: Meeting New Challenges, ASTM STP 1102, Jack R. Hall, G. Douglas Glysson, ed., American Society for Testing and Materials, Philadelphia, 1991.
6. Bej, A.K., Steffan, R.J., DiCesare, J., Haff, L., and Atlas, R.M., "Detection of Coliform Bacteria in Water by Polymerase Chain Reaction and Gene Probes", Applied and Environmental Microbiology, **56**, 306-314, 1990.
7. Bej, A.K., Mahbubani, M.H., Miller, R., DiCesare, J.L., Haff, L., and Atlas, R.M., "Multiplex PCR Amplification and Immobilized Capture Probes for Detection of Bacterial Pathogens and Indicators in Water", Molecular and Cellular Probes, **4**, 353-365, 1990.

RECOMBINANT DNA TECHNOLOGY IN THE BOVINE DISEASES CONTROL PROGRAM AND DEVELOPMENT OF CATTLE INDUSTRY

Charlya Brockelman, Vichai Boonsaeng and
Peerapan Tan-ariya

Department of Microbiology and Department of Biochemistry
Faculty of Science, Mahidol University, Rama VI Road, Bangkok
10400

ABSTRACT

For effective control of the vector-borne bovine diseases, a thorough knowledge of the nature of the infective agents, the vectorial capability, and the susceptibility of the bovine host is required. There is, however, little information on prevalence and transmission of bovine diseases due to the difficulty of identifying infected animals by conventional microscopic examination. Therefore, we have used the novel technology of diagnostic DNA-probes for detection of Babesia infection. The best probe can detect as little as 25 ng of purified Babesia DNA, the equivalent of 0.001% parasitemia. Each of the developed probes are species specific. Research on development of vaccines against other tick-borne diseases is also being pursued. The strategy is to identify surface proteins relevant to protection, clone them, and express gene coding for these proteins and test their effectiveness as protective immunogens in recombinant vaccinia construct.

INTRODUCTION

Hemoparasitic disease are endemic in half the world's cattle production areas and are the greatest constraint to improved meat and milk production in developing countries. The most wide-spread are the rickettsial disease, anaplasmosis, and the tick-borne hemoprotozoan diseases, babesiosis and theileriosis (Uilenberg, 1983). These observations have led to the initiation of the now well established Network for Research and Information on Anaplasmosis and babesiosis which operates world-wide (Burridge, 1990).

Accumulated evidence in Thailand indicates the need for a control program against tick-borne diseases, particularly in the dairy herds which consist of foreign breeds or hybrids with foreign stocks (Brockelman, 1989). In Thailand, there were attempts to produce live-attenuated vaccine by serial passage of B. bigemina infected blood through splenectomized calves (Aranyaganond et al 1975). Although the authors claimed success for their vaccine, this immunogen has never been reproduced nor used in any *Babesia* prevention program in Thailand.

One of the approaches for controlling infectious diseases occurs at the level of prevention which needs extensive data from epidemeological survey to identify endemic areas and the dynamics of disease transmission. These baseline data are also the prerequisite for vaccine trials and an immunization program. Thus, a reliable diagnostic method is important as much as an effective vaccine. With the advent of recombinant DNA technology, it is feasible to develop diagnostic DNA probes and to clone gene (s) whose products can be used as protective immunogens. This report is intended to provide an update on the implementation of this novel technology which promises to be effective and environmentally sound.

CONSTRUCTION OF BABESIA SPECIFIC DNA PROBES

For this objective, continuous culture lines of *Babesia bovis* and *B. bigemina* were established for the first time in Thailand, so that parasite DNAs devoid of host nuclear materials from white blood cells can be obtained (Brockelman and Tan-ariya (1991). Babesia DNA from each species was extracted, digested by endonuclease EcoRI, and ligated into the EcoRI site of digested plasmid, either pUN121 (Petpoo, 1990) or bluescript vector system. Plasmid vectors carrying Babesia DNA-fragments were transferred into *E. coli* JM107 and the transformed bacteria were selected on tetracycline-containing agar and screened by hybridization with ^{32}P-radiolabelled parasite DNA. The selected transformed *E. coli* colonies were amplified in tetracycline broth and isolated for plasmid clones by alkaline lysis. The clones obtained were radiolabelled with ^{32}P-ATP by nick-translation (Rigby et al, 1977) and used as hybridization probe.

Table 1. Results from comparative diagnosis of babesiosis in dairy cows using microscopic examination and diagnostic probes

Animal code	Microscopic examination			Diagnostic probes		
	B. bovis	**B.bigemina**	**B. spp.**	**B.bovis**	**B.bigemina**	**B. spp.**
ML10547	-	-	dots	-	++	
ML13747	-	-	0.8%	+	+	
ML13412	-	-	0%	-	++	
ML13894	-		0.8%	+	++	
ML11102	-		-	+	++	
ML11522	-		0.05%	-	+	
D 30015	-	10%		-	+++	
D 20150	-		dots?	-	+	
D 30018		0.03%		-	+	
Control	0.002%	-		+	-	
Control	-	0.002%		-	+	

SENSITIVITY AND SPECIFICITY OF THE PROBES

The sensitivity of the probes was determined using 2-fold dilutions of purified *Babesia* DNA. The probe was capable of detecting 50 pg of genomic DNA of the homologous species, i.e. *B. bovis* (Petpoo, 1990) or *B. bigemina*. When whole infected blood from in vitro culture was serially diluted with noninfected blood to produce varying parasite densities and tested with the probes, the probes were able to detect 300 Babesia-infected erythrocytes in 10 µl of whole blood sample. This is the equivalent of 0.001% parasitemia.

Species-specificity of the probes was confirmed using both *in vitro* culture materials and parasites in naturally infected blood from animals which has been diagnosed clinically. The capability of the probes to identify parasite species was superior over the microscopic method, in particular, for routine survey of the whole herd when the animals showed no clinical signs. In this situation, the parasite density was so low that it had to be detected by thick smears whose preparation method distorted parasite morphology. As a result, species identification was in most cases inconclusive.

Our results of comparative diagnosis of babesiosis in dairy cows using microscopic examination and diagnostic probes are summarized in Table 1. Of the nine blood samples (collected from the farm of the Dairy Promotion Organization) which gave positive results with DNA probes of either species, only two samples could be identified as *B. bigemina* by microscopic examination. By contrast, the probes differentiate all of the nine samples as follows: 3 samples were mixed-infected, and 6 samples were infected with *B. bigemina* alone. Since these results were obtained from direct dotting of the blood samples without laborious methods of processing such as extraction and purification of parasite DNA, or amplification procedures such as polymerase chain reaction, the diagnostic DNA probes for babesiosis are promising for large scale field use.

BABESIOSIS VACCINE DEVELOPMENT

Vaccination of cattle with culture derived non-virulent strains of *Babesia* to produce the stage of premunition is the most common method of immunoprophylaxis. This method has been evaluated for its biological and technical shortcomings (Burridge, 1990). To overcome these problems, novel techniques in molecular biology and immunochemistry have been employed. *B. bigemina* and *B. bovis* polypeptides on merozoite surfaces and parasitized erythrocyte surfaces-exposed epitopes were used. Four molecules: *B. bigemina* p 58, *B. bovis* 42 kD, *B. bovis* 60 kD and *B. bovis* 225 kD have been selected in developing recombinants based upon data suggesting their relevance to immune protection (McElwain, et al 1987; Hines et al, 1989). These recombinant surface proteins will be expressed in vaccinia virus. Currently, these vaccines are under development at the University of Florida.

CONCLUSION

At present, the use of diagnostic DNA probes is viewed for its application as in the initial stage economically questionable. Nevertheless, it has been agreed upon that this novel technique is uniquely able to deal with large sample size (Barker 1990). For cattle diseases in particular, the technique will allow routine examination of whole herds, detection of infection in carrier animals and tick vectors (Goff, 1988). Such data are useful in monitoring program.

Subunit vaccines obtained from genetically engineered virus will overcome some biological shortcomings, such as contamination with bovine leukemia virus as in the case of live attenuated vaccine, or acquired virulence of the non-virulent parasite strain used for immunization.

REFERENCES

Aranyaganond, P. et al 1978. Piroplasmosis vaccine. J. Thai Vet. Med. Assoc 29: 111-118.

Barker, R.H. 1990. DNA Probe diagnosis of parasitic infections. Exptl. Parasitol. 70: 494-499.

Brockelman, C.R. 1989. Prevalence and impact of anaplasmosis and babesiosis in Asia. Thai J. Trop. Med. Parasitol. 12: 31-36.

Brockelman, C.R. and Tan-ariya P. 1991. Development of an in vitro microtest to assess drug susceptibility of *Babesia bovis* and *Babesia bigemina*. J. Parasitol. 77: 994-997

Burridge, M.J. 1990. Improved control of anaplasmosis and babesiosis through biotechnology. Proc. 7th FAVA Congress, Pattaya.

Goff, W. et al. 1988. Detection of *Anaplasma marginale*-infected tick vectors by using a cloned DNA probe. Proc. Natl. Acad. Sci. (USA)85: 919-923.

Hines, S.A. et al. 1989. Molecular characterization of *Babesia bovis* merozoite surface proteins bearing epitopes immunodorminant in protected cattle. Mol. Biochem. Parasitol. 37: 1-9.

McElwain, T.F. et al. 1987. Antibodies define multiple proteins with epitopes exposed on the surface of live *Babesia bigemina* merozoites. J. Immunol. 138: 2298-2302.

Petpoo, W. 1990. Development of specific DNA probe to detect *Babesia bovis* infection in cattle.. M. Sc. Thesis Mahidol University 125 pp.

Rigby, P.W.J. et al. 1977. Labelling DNA to high specific activity in vitro by nick translation with DNA polymerase I. J. Mol. Biol. 113: 237-251.

Uilenberg, G. 1983. Epizootiology of tick-borne diseases. FAO Animal Production and Health paper. Series 36: 12-19.

SUBJECT INDEX

6 Aminopenicillanic acid, 191,196
Agrobacterium tumefaciens, 65, 55-61
Alcaligenes eutrophus , 110-111, 114, 116-118
Alginate,102-103
alp gene, 206
Amino acid sequence, 68
Arabidopsis thaliana , 66-67
Arsenate resistance, 114-116
Azotobacter , 116

B. megaterium , 191,194
B. sterothermophilus , 201-203
B. thuringiensis , 179-187
Babesia , 225-228
Babesiosis, 225, 227
Bacillus, 120, 165-7, 175-7, 179-187, 191-6, 201-203,
    206-209
Bacterial blight, *see Xanthomonas*
Bacterial colonization, 86-87
Bacterial immobilization, 196
Bioremendation, 103-106, 131
Biotechnology, 76
Bovine diseases, 225
Brassica , 65-68

Cadmium resistance, 110-112
Candida maltosa , 13-22
Candida maltosa plasmid, 13-22
CD4 cells, 4-6
Chloramphenicol, 165-7
Chloramphenicol acetyltransferase (CAT), 28-29,
    165-167, 175-177
Chlorinated hydrocarbon, 143-146
Cholera, 211-213
Chromate resistance, 116-118
Coat protein, 49-53
Cometabolic substrates, 143-144
Complementation, 17
Conjugation like process, 179
Copper resistance, 112-114
Crown gall tumor, 55

αA-CRBP1, 27-30
Crystallin gene, 27-28
Crystallization, 197-198
Cystic fibrosis, 100-102

Dehalogenation oxidative, 145-146
Dehalogenation reductive, 146
Dehalogenation substituitive, 144-145
Detoxification, 109-124
2,4-dichlorophenoxyacetic acid, 103-105
DNA probes, 91-94, 219-222, 225-227
Double strand gap repair, 5-6

Electron acceptors, 143
Embryonic chicken fibroblast, 28
Enterobacter cloacae, 117-118
Enterotoxins, 211
Enzyme evolution, 149, 152
Episomal DNA, 3-7
Epitopes, 211, 215-217
Eukaryotic moveable elements, 7-10
Evolution, 99

Fibroblast, 28-29
Field trials, 52
Finger prints, 96
Flavobacterium sp., 149

Gene probes, 219-223, 225, 227
Gene replacement, 170
Gene expression:
    bacterial, 58-62, 100-106, 139-142, 155-161, 165-167,
        175-177
    eukaryotic, 7-10, 13-22, 23-25, 27-31, 38-44
Gene revolution, 71
Genetic mapping, 171
β-1,2-glucan, 56
β-1,3-glucanase, 23-26
β-glucuronidase (GUS), 65
Genomic libraries, 170
Gramineae, 94, *see* rice

229